Lecture Notes in Mathematics

Edited by J.-M. Morel, F. Takens and B. Teissier

Editorial Policy
for the publication of monographs

1. Lecture Notes aim to report new developments in all areas of mathematics and their applications – quickly, informally and at a high level. Mathematical texts analysing new developments in modelling and numerical simulation are welcome.

   Monograph manuscripts should be reasonably self-contained and rounded off. Thus they may, and often will, present not only results of the author but also related work by other people. They may be based on specialised lecture courses. Furthermore, the manuscripts should provide sufficient motivation, examples and applications. This clearly distinguishes Lecture Notes from journal articles or technical reports which normally are very concise. Articles intended for a journal but too long to be accepted by most journals, usually do not have this „lecture notes" character. For similar reasons it is unusual for doctoral theses to be accepted for the Lecture Notes series, though habilitation theses may be appropriate.

2. Manuscripts should be submitted (preferably in duplicate) either to Springer's mathematics editorial in Heidelberg, or to one of the series editors (with a copy to Springer). In general, manuscripts will be sent out to 2 external referees for evaluation. If a decision cannot yet be reached on the basis of the first 2 reports, further referees may be contacted: The author will be informed of this. A final decision to publish can be made only on the basis of the complete manuscript, however a refereeing process leading to a preliminary decision can be based on a pre-final or incomplete manuscript. The strict minimum amount of material that will be considered should include a detailed outline describing the planned contents of each chapter, a bibliography and several sample chapters.

   Authors should be aware that incomplete or insufficiently close to final manuscripts almost always result in longer refereeing times and nevertheless unclear referees' recommendations, making further refereeing of a final draft necessary.

   Authors should also be aware that parallel submission of their manuscript to another publisher while under consideration for LNM will in general lead to immediate rejection.

3. Manuscripts should in general be submitted in English. Final manuscripts should contain at least 100 pages of mathematical text and should always include

   – a table of contents;
   – an informative introduction, with adequate motivation and perhaps some historical remarks: it should be accessible to a reader not intimately familiar with the topic treated;
   – a subject index: as a rule this is genuinely helpful for the reader.

   For evaluation purposes, manuscripts may be submitted in print or electronic form (print form is still preferred by most referees), in the latter case preferably as pdf- or zipped ps-files. Lecture Notes volumes are, as a rule, printed digitally from the authors' files. To ensure best results, authors are asked to use the LaTeX2e style files available from Springer's web-server at:

   ftp://ftp.springer.de/pub/tex/latex/mathegl/mono/ (for monographs) and

   ftp://ftp.springer.de/pub/tex/latex/mathegl/mult/ (for summer schools/tutorials).

   Additional technical instructions, if necessary, are available on request from lnm@springer-sbm.com.

Continued on inside back-cover

Lecture Notes in Mathematics 1920

Editors:
J.-M. Morel, Cachan
F. Takens, Groningen
B. Teissier, Paris

Subseries:
École d'Été de Probabilités de Saint-Flour

Saint-Flour Probability Summer School

The Saint-Flour volumes are reflections of the courses given at the Saint-Flour Probability Summer School. Founded in 1971, this school is organised every year by the Laboratoire de Mathématiques (CNRS and Université Blaise Pascal, Clermont-Ferrand, France). It is intended for PhD students, teachers and researchers who are interested in probability theory, statistics, and in their applications.

The duration of each school is 13 days (it was 17 days up to 2005), and up to 70 participants can attend it. The aim is to provide, in three high-level courses, a comprehensive study of some fields in probability theory or Statistics. The lecturers are chosen by an international scientific board. The participants themselves also have the opportunity to give short lectures about their research work.

Participants are lodged and work in the same building, a former seminary built in the 18th century in the city of Saint-Flour, at an altitude of 900 m. The pleasant surroundings facilitate scientific discussion and exchange.

The Saint-Flour Probability Summer School is supported by:

– Université Blaise Pascal
– Centre National de la Recherche Scientifique (C.N.R.S.)
– Ministère délégué à l'Enseignement supérieur et à la Recherche

For more information, see back pages of the book and
http://math.univ-bpclermont.fr/stflour/

Jean Picard
Summer School Chairman
Laboratoire de Mathématiques
Université Blaise Pascal
63177 Aubière Cedex
France

Steven N. Evans

Probability and Real Trees

École d'Été de Probabilités
de Saint-Flour XXXV - 2005

 Springer

Author

Steven Neil Evans
Department of Statistics # 3860
367 Evans Hall
University of California at Berkeley
Berkeley, CA 94720-3860
USA
e-mail: evans@stat.Berkeley.EDU
URL: http://www.stat.berkeley.edu/users/evans

Cover: Blaise Pascal (1623-1662)

Library of Congress Control Number: 2007934014

Mathematics Subject Classification (2000): 60B99, 05C05, 51F99, 60J25

ISSN print edition: 0075-8434
ISSN electronic edition: 1617-9692
ISSN Ecole d'Eté de Probabilités de St. Flour, print edition: 0721-5363
ISBN 978-3-540-74797-0 Springer Berlin Heidelberg New York
DOI 10.1007/978-3-540-74798-7

This work is subject to copyright. All rights are reserved, whether the whole or part of the material is concerned, specifically the rights of translation, reprinting, reuse of illustrations, recitation, broadcasting, reproduction on microfilm or in any other way, and storage in data banks. Duplication of this publication or parts thereof is permitted only under the provisions of the German Copyright Law of September 9, 1965, in its current version, and permission for use must always be obtained from Springer. Violations are liable for prosecution under the German Copyright Law.

Springer is a part of Springer Science+Business Media
springer.com
© Springer-Verlag Berlin Heidelberg 2008

The use of general descriptive names, registered names, trademarks, etc. in this publication does not imply, even in the absence of a specific statement, that such names are exempt from the relevant protective laws and regulations and therefore free for general use.

Typesetting by the author and SPi using a Springer LaTeX macro package

Cover art: Tree of Life, with kind permission of David M. Hillis, Derrick Zwickl, and Robin Gutell, University of Texas.

Cover design: WMX Design, Heidelberg

Printed on acid-free paper   SPIN: 12114894   VA41/3100/SPi   5 4 3 2 1 0

For Ailan Hywel, Ciaran Leuel and Huw Rhys

Preface

These are notes from a series of ten lectures given at the Saint–Flour Probability Summer School, July 6 – July 23, 2005.

The research that led to much of what is in the notes was supported in part by the U.S. National Science Foundation, most recently by grant DMS-0405778, and by a Miller Institute for Basic Research in Science Research Professorship.

Some parts of these notes were written during a visit to the Pacific Institute for the Mathematical Sciences in Vancouver, Canada. I thank my long-time collaborator Ed Perkins for organizing that visit and for his hospitality. Other portions appeared in a graduate course I taught in Fall 2004 at Berkeley. I thank Rui Dong for typing up that material and the students who took the course for many useful comments. Judy Evans, Richard Liang, Ron Peled, Peter Ralph, Beth Slikas, Allan Sly and David Steinsaltz kindly proof-read various parts of the manuscript.

I am very grateful to Jean Picard for all his work in organizing the Saint–Flour Summer School and to the other participants of the School, particularly Christophe Leuridan, Cedric Villani and Matthias Winkel, for their interest in my lectures and their suggestions for improving the notes.

I particularly acknowledge my wonderful collaborators over the years whose work with me appears here in some form: David Aldous, Martin Barlow, Peter Donnelly, Klaus Fleischmann, Tom Kurtz, Jim Pitman, Richard Sowers, Anita Winter, and Xiaowen Zhou. Lastly, I thank my friend and collaborator Persi Diaconis for advice on what to include in these notes.

Berkeley, California, U.S.A.

Steven N. Evans
October 2006

Contents

1

Introduction

The *Oxford English Dictionary* provides the following two related definitions of the word *phylogeny*:

1. The pattern of historical relationships between species or other groups resulting from divergence during evolution.
2. A diagram or theoretical model of the sequence of evolutionary divergence of species or other groups of organisms from their common ancestors.

In short, a phylogeny is the "family tree" of a collection of units designated generically as *taxa*. Figure 1.1 is a simple example of a phylogeny for four primate species. Strictly speaking, phylogenies need not be trees. For instance, biological phenomena such as hybridization and horizontal gene transfer can lead to non-tree-like *reticulate* phylogenies for organisms. However, we will only be concerned with trees in these notes.

Phylogenetics (that is, the construction of phylogenies) is now a huge enterprise in biology, with several sophisticated computer packages employed extensively by researchers using massive amounts of DNA sequence data to study all manner of organisms. An introduction to the subject that is accessible to mathematicians is [67], while many of the more mathematical aspects are surveyed in [125].

It is often remarked that a tree is the only illustration Charles Darwin included in *The Origin of Species*. What is less commonly noted is that Darwin acknowledged the prior use of trees as representations of evolutionary relationships in historical linguistics – see Figure 1.2. A recent collection of papers on the application of computational phylogenetic methods to historical linguistics is [69].

The diversity of life is enormous. As J.B.S Haldane often remarked[1] in various forms:

---

[1] See Stephen Jay Gould's essay "A special fondness for beetles" in his book [77] for a discussion of the occasions on which Haldane may or may not have made this remark.

Fig. 1.1. The phylogeny of four primate species. Illustrations are from the *Tree of Life Web Project* at the University of Arizona

I don't know if there is a God, but if He exists He must be inordinately fond of beetles.

Thus, phylogenetics leads naturally to the consideration of very large trees – see Figure 1.3 for a representation of what the phylogeny of all organisms might look like and browse the Tree of Life Web Project web-site at `http://www.tolweb.org/tree/` to get a feeling for just how large the phylogenies of even quite specific groups (for example, beetles) can be.

Not only can phylogenetic trees be very large, but the number of possible phylogenetic trees for even a moderate number of taxa is enormous. Phylogenetic trees are typically thought of as rooted bifurcating trees with only the leaves labeled, and the number of such trees for n leaves is $(2n-3) \times (2n-5) \times \cdots \times 7 \times 5 \times 3 \times 1$ – see, for example, Chapter 3 of [67]. Consequently, if we try to use statistical methods to find the "best" tree that fits a given set of data, then it is impossible to exhaustively search all possible trees and we must use techniques such as Bayesian Markov Chain Monte Carlo and simulated annealing that randomly explore tree space in some way. Hence phylogenetics leads naturally to the study of large random trees and stochastic processes that move around spaces of large trees.

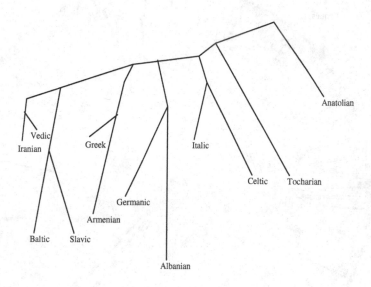

Fig. 1.2. One possible phylogenetic tree for the Indo-European family of languages from [118]

Although the investigation of random trees has a long history stretching back to the eponymous work of Galton and Watson on branching processes, a watershed in the area was the sequence of papers by Aldous [12, 13, 10]. Previous authors had considered the asymptotic behavior of numerical features of an ensemble of random trees such as their height, total number of vertices, average branching degree, etc. Aldous made sense of the idea of a sequence of trees converging to a limiting "tree-like object", so that many such limit results could be read off immediately in a manner similar to the way that limit theorems for sums of independent random variables are straightforward consequences of Donsker's invariance principle and known properties of Brownian motion. Moreover, Aldous showed that, akin to Donsker's invariance principle, many different sequences of random trees have the same limit, the *Brownian continuum random tree*, and that this limit is essentially the standard Brownian excursion "in disguise".

We briefly survey Aldous's work in Chapter 2, where we also present some of the historical development that appears to have led up to it, namely the probabilistic proof of the Markov chain tree theorem from [21] and the algorithm of [17, 35] for generating uniform random trees that was inspired by that proof. Moreover, the asymptotic behavior of the tree–generating algorithm

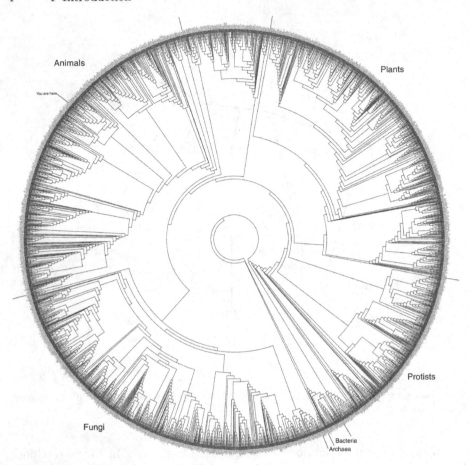

Fig. 1.3. A somewhat impressionistic depiction of the phylogenetic tree of all life produced by David M. Hillis, Derrick Zwickl, and Robin Gutell, University of Texas

when the number of vertices is large is the subject of Chapter 5, which is based on [63].

Perhaps the key conceptual difficulty that Aldous had to overcome was how to embed the collection of finite trees into a larger universe of "tree-like objects" that can arise as re-scaling limits when the number of vertices goes to infinity. Aldous proposed two devices for doing this. Firstly, he began with a classical bijection, due to Dyck, between rooted planar trees and suitable lattice paths (more precisely, the sort of paths that can appear as the "positive excursions" of a simple random walk). He showed how such an encoding of trees as continuous functions enables us to make sense of weak convergence of random trees as just weak convergence of random functions (in the sense of weak convergence with respect to the usual supremum norm). Secondly, he noted that a finite tree with edge lengths is naturally isomorphic to a compact subset of ℓ^1, the space of absolutely summable sequences. This enabled him to treat weak convergence of random trees as just weak convergence of random compact sets (where compact subsets of ℓ^1 are equipped with the Hausdorff distance arising from the usual norm on ℓ^1).

Although Aldous's approaches are extremely powerful, the identification of trees as continuous functions or compact subsets of ℓ^1 requires, respectively, that they are embedded in the plane or leaf-labeled. This embedding or labeling can be something of an artifact when the trees we are dealing with don't naturally come with such a structure. It can be particularly cumbersome when we are considering tree-valued stochastic processes, where we have to keep updating an artificial embedding or labeling as the process evolves. Aldous's perspective is analogous to the use of coordinates in differential geometry: explicit coordinates are extremely useful for many calculations but they may not always offer the smoothest approach. Moreover, it is not clear *a priori* that every object we might legitimately think of as tree-like necessarily has a representation as an excursion path or a subset of ℓ^1. Also, the topologies inherited from the supremum norm or the Hausdorff metric may be too strong for some purposes.

We must, therefore, seek more intrinsic ways of characterizing what is meant by a "tree-like object". Finite combinatorial trees are just graphs that are connected and acyclic. If we regard the edges of such a tree as intervals, so that a tree is a cell complex (and, hence, a particular type of topological space), then these two defining properties correspond respectively to connectedness in the usual topological sense and the absence of subspaces that are homeomorphic to the circle. Alternatively, a finite combinatorial tree thought of as a cell complex has a natural metric on it: the distance between two points is just the length of the unique "path" through the tree connecting them (where each edge is given unit length). There is a well-known characterization of the metrics that are associated with trees that is often called *(Buneman's) four point condition* – see Chapter 3. Its significance seems to have been recognized independently in [149, 130, 36] – see [125] for a discussion of the history.

These observations suggest that the appropriate definition of a "tree-like object" should be a general topological or metric space with analogous properties. Such spaces are called $\mathbb{R}$-*trees* and they have been studied extensively – see [46, 45, 137, 39]. We review some of the relevant theory and the connection with 0-*hyperbolicity* (which is closely related to the four point condition) in Chapter 3.

We note in passing that $\mathbb{R}$-trees, albeit ones with high degrees of symmetry, play an important role in geometric group theory – see, for example, [126, 110, 127, 30, 39]. Also, 0-hyperbolic metric spaces are the simplest example of the δ-hyperbolic metric spaces that were introduced in [79] as a class of spaces with global features similar to those of complete, simply connected manifolds of negative curvature. For more on the motivation and subsequent history of this notion, we refer the reader to [33, 39, 80]. Groups with a natural

δ-hyperbolic metric have turned out to be particularly important in a number of areas of mathematics, see [79, 20, 40, 76].

In order to have a nice theory of random $\mathbb{R}$-trees and $\mathbb{R}$-tree-valued stochastic processes, it is necessary to metrize a collection of $\mathbb{R}$-trees, and, since $\mathbb{R}$-trees are just metric spaces with certain special properties, this means that we need a way of assigning a distance between two metric spaces (or, more correctly, between two isometry classes of metric spaces). The *Gromov-Hausdorff distance* – see [80, 37, 34] – does exactly this and turns out to be very pleasant to work with. The particular properties of the Gromov-Hausdorff distance for collections of $\mathbb{R}$-trees have been investigated in [63, 65, 78] and we describe some of the resulting theory in Chapter 4.

Since we introduced the idea of using the formalism of $\mathbb{R}$-trees equipped with the Gromov-Hausdorff metric to study the asymptotics of large random trees and tree-valued processes in [63, 65], there have been several papers that have adopted a similar point of view – see, for example, [49, 101, 102, 103, 50, 81, 78].

As we noted above, stochastic processes that move through a space of finite trees are an important ingredient for several algorithms in phylogenetic analysis. Usually, such chains are based on a set of simple rearrangements that transform a tree into a "neighboring" tree. One standard set of moves that is implemented in several phylogenetic software packages is the set of *subtree prune and re-graft* (SPR) moves that were first described in [134] and are further discussed in [67, 19, 125]. Moreover, as remarked in [19],

> The SPR operation is of particular interest as it can be used to model biological processes such as horizontal gene transfer and recombination.

Section 2.7 of [125] provides more background on this point as well as a comment on the role of SPR moves in the two phenomena of lineage sorting and gene duplication and loss. Following [65], we investigate in Chapter 9 the behavior when the number of vertices goes to infinity of the simplest Markov chain based on SPR moves.

Tree-valued Markov processes appear in contexts other than phylogenetics. For example, a number of such processes appear in combinatorics associated with the random graph process, stochastic coalescence, and spanning trees – see [115]. One such process is the *wild chain*, a Markov process that appears as a limiting case of tree–valued Markov chains arising from pruning operations on Galton–Watson and conditioned Galton–Watson trees in [16, 14].

The state space of the wild chain is the set $\mathbf{T}^*$ consisting of rooted $\mathbb{R}$-trees such that each edge has length 1, each vertex has finite degree, and if the tree is infinite there is a single path of infinite length from the root. The wild chain is reversible (that is, symmetric). Its equilibrium measure is the distribution of the critical Poisson Galton–Watson branching process (we denote this probability measure on rooted trees by PGW(1)). When started in a state that is a finite tree, the wild chain holds for an exponentially distributed amount of time and then jumps to a state that is an infinite tree. Then, as must be the case given that the PGW(1) distribution assigns all of its mass to finite trees, the process instantaneously re-enters the set of finite trees. In other words, the sample–paths of the wild chain bounce backwards and forwards between the finite and infinite trees.

As we show in Chapter 6 following [15], the wild chain is a particular instance of a general class of symmetric Markov processes that spend Lebesgue almost all of their time in a countable, discrete part of their state-space but continually bounce back and forth between this region and a continuous "boundary". Other processes in this general class are closely related to the Markov processes on totally disconnected Abelian groups considered in [59]. A special case of these latter group-valued processes, where the group is the additive group of a local field such as the p-adic numbers, is investigated in [4, 5, 7, 6, 2, 8, 9, 87, 131, 68].

Besides branching models such as Galton–Watson processes, another familiar source of random trees is the general class of coalescing models – see [18] for a recent survey and bibliography.

Kingman's coalescent was introduced in [90, 89] as a model for genealogies in the context of population genetics and has since been the subject of a large amount of applied and theoretical work – see [136, 144, 83] for an indication of some of the applications of Kingman's coalescent in genetics.

Families of coalescing Markov processes appear as duals to interacting particle systems such as the voter model and stepping stone models . Motivated by this connection, [22] investigated systems of coalescing Brownian motions and the closely related coalescing Brownian flow . Coalescing Brownian motion has recently become a topic of renewed interest, primarily in the study of filtrations and "noises" – see, for example, [140, 132, 138, 55].

In Chapter 8 we show, following [60, 44], how Kingman's coalescent and systems of coalescing Brownian motions on the circle are each naturally associated with random compact metric spaces and we investigate the fractal properties of those spaces. A similar study was performed in [28] for trees arising from the beta-coalescents of [116]. There has been quite a bit of work on fractal properties of random trees constructed in various ways from Galton–Watson branching processes; for example, [82] computed the Hausdorff dimension of the boundary of a Galton–Watson tree equipped with a natural metric – see also [104, 96].

We observe that Markov processes with *continuous* sample paths that take values in a space of continuous excursion paths and are reversible with respect to the distribution of standard Brownian excursion have been investigated in [148, 147, 146]. These processes can be thought of as $\mathbb{R}$-tree valued diffusion processes that are reversible with respect to the distribution of the Brownian continuum random tree.

Moving in a slightly different but related direction, there is a large literature on random walks with state-space a given infinite tree: [145, 105] are excellent bibliographical references. In particular, there is a substantial amount of research on the Martin boundary of such walks beginning with [52, 38, 122].

The literature on diffusions on tree–like or graph–like structures is more modest. A general construction of diffusions on graphs using Dirichlet form methods is given in [141]. Diffusions on tree–like objects are studied in [42, 93] using excursion theory ideas, local times of diffusions on graphs are investigated in [53, 54], and an averaging principle for such processes is considered in [71]. One particular process that has received a substantial amount of attention is the so-called *Walsh's spider*. The spider is a diffusion on the tree consisting of a finite number of semi–infinite rays emanating from a single vertex – see [142, 26, 139, 25].

A higher dimensional diffusion with a structure somewhat akin to that of the spider, in which regions of higher dimensional spaces are "glued" together along lower dimensional boundaries, appears in the work of Sowers [133] on Hamiltonian systems perturbed by noise – see also [111]. A general construction encompassing such processes is given in [64]. This construction was used in [24] to build diffusions on the interesting fractals introduced in [95] to answer a question posed in [84].

In Chapter 7 we describe a particular Markov process with state–space an $\mathbb{R}$-tree that does not have any leaves (in the sense that any path in the tree can be continued indefinitely in both directions). The initial study of this process in [61] was motivated by Le Gall's *Brownian snake* process – see, for example, [97, 98, 99, 100]. One agreeable feature of this process is that it serves as a new and convenient "test bed" on which we can study many of the objects of general Markov process theory such as Doob h-transforms, the classification of entrance laws, the identification of the Martin boundary and representation of excessive functions, and the existence of non-constant harmonic functions and the triviality of tail σ-fields.

We use Dirichlet form methods in several chapters, so we have provided a brief summary of some of the more salient parts of the theory in Appendix A. Similarly, we summarize some results on Hausdorff dimension, packing dimension and capacity that we use in various places in Appendix B.

2

Around the Continuum Random Tree

2.1 Random Trees from Random Walks

2.1.1 Markov Chain Tree Theorem

Suppose that we have a discrete time Markov chain $X = \{X_n\}_{n \in \mathbb{N}_0}$ with state space V and irreducible transition matrix P. Let π be the corresponding stationary distribution. The *Markov chain tree theorem* gives an explicit formula for π, as opposed to the usual implicit description of π as the unique probability vector that solves the equation $\pi P = \pi$. In order to describe this result, we need to introduce some more notation.

Let $G = (V, E)$ be the directed graph with vertex set V and directed edges consisting of pairs of vertices (i, j) such that $p_{ij} > 0$. We call p_{ij} the *weight* of the edge (i, j).

A *rooted spanning tree of* G is a directed subgraph of G that is a spanning tree as an undirected graph (that is, it is a connected subgraph without any cycles that has V as its vertex set) and is such that each vertex has out-degree 1, except for a distinguished vertex, the *root*, that has out-degree 0. Write $\mathcal{A}$ for the set of all rooted spanning trees of G and $\mathcal{A}_i$ for the set of rooted spanning trees that have i as their root.

The *weight* of a rooted spanning tree T is the product of its edge weights, which we write as weight(T).

Theorem 2.1. *The stationary distribution* π *is given by*

$$\pi_i = \frac{\sum_{T \in \mathcal{A}_i} \text{weight}(T)}{\sum_{T \in \mathcal{A}} \text{weight}(T)}.$$

Proof. Let $\bar{X} = \{\bar{X}_n\}_{n \in \mathbb{Z}}$ be a two-sided stationary Markov chain with the transition matrix P (so that $\bar{X}_n$ has distribution π for all $n \in \mathbb{Z}$).

Define a map $f : V^{\mathbb{Z}} \to \mathcal{A}$ as follows – see Figure 2.2.

- The root of $f(x)$ is x_0.

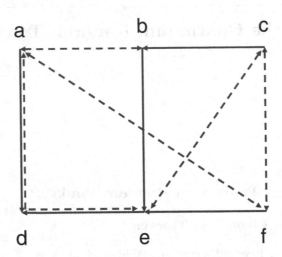

Fig. 2.1. A rooted spanning tree. The solid directed edges are in the tree, whereas dashed directed edges are edges in the underlying graph that are not in the tree. The tree is rooted at d. The weight of this tree is $p_{ad}p_{ed}p_{be}p_{fe}p_{cb}$.

- For $i \neq x_0$, the unique edge in $f(x)$ with tail i is $(i, x_{\tau(i)+1}) = (x_{\tau(i)}, x_{\tau(i)+1})$ where $\tau(i) := \sup\{m < 0 : x_m = i\}$.

It is clear that f is well-defined almost surely under the distribution of $\bar{X}$ and so we can define a stationary, $\mathcal{A}$-valued, $\mathbb{Z}$-indexed stochastic process $\bar{Y} = \{\bar{Y}_n\}_{n \in \mathbb{Z}}$ by

$$\bar{Y}_n := f(\theta^n(\bar{X})), \quad n \in \mathbb{Z},$$

where $\theta : V^{\mathbb{Z}} \to V^{\mathbb{Z}}$ denotes the usual shift operator defined by $\theta(x)_n := x_{n+1}$.

It is not hard to see that $\bar{Y}$ is Markov. More specifically, consider the following *forward procedure* that produces a spanning tree rooted at j from a spanning tree S rooted at i – see Figure 2.3.

- Attach the directed edge (i, j) to S.
- This creates a directed graph with unique directed loop that contains i and j (possibly a self loop at i).
- Delete the unique directed edge out of j.
- This deletion breaks the loop and produces a spanning tree rooted at j.

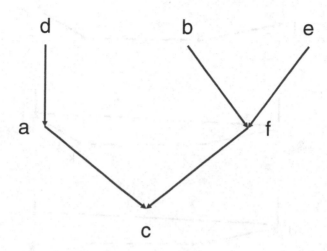

Fig. 2.2. The construction of the rooted tree $f(x)$ for $V = \{a, b, c, d, e, f\}$ and $(\dots, x_{-2}, x_{-1}, x_0) = (\dots, e, f, c, a, c, d, d, a, f, b, f, a, c, c, f, c)$

Then, given $\{\bar{Y}_m : m \leqslant n\}$, the tree $\bar{Y}_{n+1}$ is obtained from the tree $\bar{Y}_n$ with root i by choosing the new root j in the forward procedure with conditional probability p_{ij}.

It is easy to see that a rooted spanning tree $T \in \mathcal{A}$ can be constructed from $S \in \mathcal{A}$ by the forward procedure if and only if S can be constructed from T by the following *reverse procedure* for a suitable vertex k.

- Let T have root j.
- Attach the directed edge (j, k) to T.
- This creates a directed graph with unique directed loop containing j and k (possibly a self loop at j).
- Delete the unique edge, say (i, j), directed into j that lies in this loop.
- This deletion breaks the loop and produces a rooted spanning tree rooted at i.

Moving up rather than down the page in Figure 2.3 illustrates the reverse procedure.

Let S and T be rooted spanning trees such that T can be obtained from S by the forward procedure, or, equivalently, such that S can be obtained from T by the reverse procedure. Write i and j for the roots of S and T,

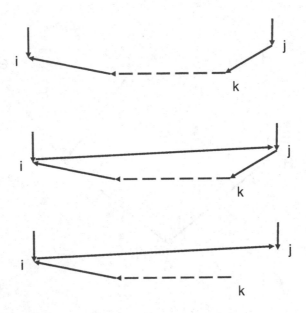

Fig. 2.3. The forward procedure. The dashed line represents a directed path through the tree that may consist of several directed edges.

respectively, and write k for the (unique) vertex appearing in the description of the reverse procedure. Denote by Q the transition matrix of the $\mathcal{A}$-valued process $\bar{Y}$. We have observed that

- If S has root i and T has root j, then $Q_{ST} = p_{ij}$.
- To get T from S we first attached the edge (i, j) and then deleted the unique outgoing edge (j, k) from j.
- To get S from T we would attach the edge (j, k) to T and then delete the edge (i, j).

Thus, if we let ρ be the probability measure on $\mathcal{A}$ such that ρ_U is proportional to the weight of U for $U \in \mathcal{A}$, then we have

$$\rho_S Q_{ST} = \rho_T R_{TS},$$

where $R_{TS} := p_{jk}$. In particular,

$$\sum_S \rho_S Q_{ST} = \sum_S \rho_T R_{TS} = \rho_T,$$

since R is a stochastic matrix. Hence ρ is the stationary distribution corresponding to the irreducible transition matrix Q. That is, ρ is the one-dimensional marginal of the stationary chain $\bar{Y}$. We also note in passing that R is the transition matrix of the time-reversal of $\bar{Y}$.

Thus,

$$\pi_i = \sum \{\rho_T : \text{ root of } T \text{ is } i\}$$
$$= \frac{\sum_{T \in \mathcal{A}_i} \text{weight}(T)}{\sum_{T \in \mathcal{A}} \text{weight}(T)},$$

as claimed.

□

The proof we have given of Theorem 2.1 is from [21], where there is a discussion of the history of the result.

2.1.2 Generating Uniform Random Trees

Proposition 2.2. *Let $(X_j)_{j \in \mathbb{N}_0}$ be the natural random walk on the complete graph K_n with transition matrix P given by $P_{ij} := \frac{1}{n-1}$ for $i \neq j$ and X_0 uniformly distributed. Write*

$$\tau_\nu = \min\{j \geq 0 : X_j = \nu\}, \quad \nu = 1, 2, \ldots, n.$$

Let T be the directed subgraph of K_n with edges

$$(X_{\tau_\nu}, X_{\tau_\nu - 1}), \quad \nu \neq X_0.$$

Then T is uniformly distributed over the rooted spanning trees of K_n.

Proof. The argument in the proof of Theorem 2.1 plus the time-reversibility of X.

□

Remark 2.3. The set of rooted spanning trees of the complete graph K_n is just the set of of n^{n-1} rooted trees with vertices labeled by $\{1, 2, \ldots, n\}$, and so the random tree T produced in Proposition 2.2 is nothing other than a uniform rooted random tree with n labeled vertices.

Proposition 2.2 suggests a procedure for generating uniformly distributed rooted random trees with n labeled vertices. The most obvious thing to do would be to run the chain X until all n states had appeared and then construct the tree T from the resulting sample path. The following algorithm, presented independently in [17, 35], improves on this naive approach by, in effect, generating X_0 and the pairs $(X_{\tau_\nu - 1}, X_{\tau_\nu})$, $\nu \neq X_0$ without generating the rest of the sample path of X.

Algorithm 2.4. *Fix $n \geq 2$. Let $U_2, \ldots, U_n$ be independent and uniformly distributed on $1, \ldots, n$, and let Π be an independent uniform random permutation of $1, \ldots, n$.*

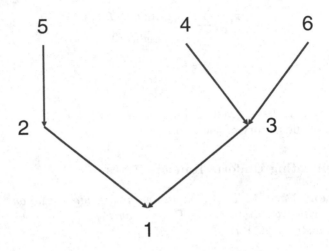

Fig. 2.4. Step (i) of Algorithm 2.4 for $n = 6$ and $(V_2, V_3, V_4, V_5, V_6) = (1, 1, 3, 2, 3)$

*(i) For $2 \leqslant i \leqslant n$ connect vertex i to vertex $V_i = (i - 1) \wedge U_i$ (that is, build a
tree rooted at 1 with edges (i, V_i)).*
(ii)Relabel the vertices $1, \dots, n$ as $\Pi_1, \dots, \Pi_n$ to produce a tree rooted at Π_1.

See Figure 2.4 for an example of Step (i) of the algorithm.

Proposition 2.5. *The rooted random tree with n labeled vertices produced by
Algorithm 2.4 is uniformly distributed.*

Proof. Let $Z_0, Z_1, \dots$ be independent and uniform on $1, 2, \dots, n$. Define
$\pi_1, \pi_2, \dots, \pi_n, \xi_1, \xi_2, \dots, \xi_n \in \mathbb{N}_0$ and $\lambda_2, \dots, \lambda_n \in \{1, 2, \dots, n\}$ by

$$\xi_1 := 0,$$

$$\pi_1 := Z_0,$$

$$\xi_{i+1} := \min\{m > \xi_i : Z_m \notin \{\pi_1, \dots, \pi_i\}\}, \quad 1 \leqslant i \leqslant n - 1,$$

$$\pi_{i+1} := Z_{\xi_{i+1}}, \quad 1 \leqslant i \leqslant n - 1,$$

$$\lambda_{i+1} := Z_{\xi_{i+1}-1}, \quad 1 \leqslant i \leqslant n - 1.$$

Consider the random tree T labeled by $\{1, 2, \dots, n\}$ with edges (π_i, λ_i), $2 \leqslant
i \leqslant n$.

Note that this construction would give the same tree if the sequence Z was replaced by the subsequence Z' in which terms Z_i identical with their predecessor Z_{i-1} were deleted. The process Z' is just the natural random walk on the complete graph. Thus, the construction coincides with the construction of Proposition 2.2. Hence, the tree T is a uniformly distributed tree on n labeled vertices. To complete the proof, we need only argue that this construction is equivalent to Algorithm 2.4.

It is clear that π is a uniform random permutation. The construction of the tree of T can be broken into two stages.

(i) Connect i to $\pi_{\lambda_i}^{-1}$, $i = 2, \ldots, n$.
(ii) Relabel $1, \ldots, n$ as $\pi_1, \ldots, \pi_n$.

Thus, it will suffice to show that the conditional joint distribution of the random variables $\pi_{\lambda_i}^{-1}$, $i = 2, \ldots, n$, given π is always the same as the (unconditional) joint distribution of the random variables V_i, $i = 2, \ldots, n$, in Algorithm 2.4 no matter what the value of π is.

To see that this is so, first fix i and condition on $Z_1, \ldots, Z_{\xi_i}$ as well as π. Note the following two facts.

- With probability $1 - i/n$ we have $\xi_{i+1} = \xi_i + 1$, which implies that $\lambda_{i+1} = Z_{\xi_i}$ and, hence, $\pi_{\lambda_{i+1}}^{-1} = i$.
- Otherwise, $\xi_{i+1} = \xi_i + M + 1$ for some random integer $M \geqslant 1$. Conditioning on the event $\{M = m\}$, we have that the random variables $Z_{\xi_i+1}, \ldots, Z_{\xi_i+m}$ are independent and uniformly distributed on the previously visited states $\{\pi_1, \ldots, \pi_i\}$. In particular, $\lambda_{i+1} = Z_{\xi_i+m}$ is uniformly distributed on $\{\pi_1, \ldots, \pi_i\}$, and so $\pi_{\lambda_{i+1}}^{-1}$ is uniformly distributed on $\{1, \ldots, i\}$.

Combining these two facts, we see that

$$\mathbb{P}\{\pi_{\lambda_{i+1}}^{-1} = u \mid Z_1, \ldots, Z_{\xi_i}, \pi\} = 1/n = \mathbb{P}\{V_{i+1} = u\}, \quad 1 \leqslant u \leqslant i - 1,$$

$$\mathbb{P}\{\pi_{\lambda_{i+1}}^{-1} = i \mid Z_1, \ldots, Z_{\xi_i}, \pi\} = 1 - (i-1)/n = \mathbb{P}\{V_{i+1} = i\},$$

as required.   □

2.2 Random Trees from Conditioned Branching Processes

If we were to ask most probabilists to propose a natural model for generating random trees, they would first think of the *family tree* of a Galton–Watson branching process. Such a tree has a random number of vertices and if we further required that the random tree had a fixed number n of vertices, then they would suggest simply conditioning the total number of vertices in the Galton-Watson tree to be n. Interestingly, special cases of this mechanism for generating random trees produce trees that are also natural from a combinatorial perspective, as we shall soon see.

Let $(p_i)_{i \in \mathbb{N}_0}$ be a probability distribution on the non-negative integers that has mean one. Write T for the family tree of the Galton–Watson branching process with offspring distribution $(p_i)_{i \in \mathbb{N}_0}$ started with 1 individual in generation 0. For $n \geq 1$ denote by T_n a random tree that arises by conditioning on the total population size $|T|$ being n (we suppose that the event $\{|T| = n\}$ has positive probability). More precisely, we think of the trees T and T_n as rooted *ordered* trees: a rooted tree is *ordered* if we distinguish the offspring of a vertex according with a "birth order". Equivalently, a rooted ordered tree is a rooted *planar* tree: the birth order is given by the left-to-right ordering of offspring in the given embedding of the tree in the plane. The distribution of the random tree T is then

$$\mathbb{P}\{T = t\} = \prod_{v \in t} p_{d(v,t)}$$

$$= \prod_{i \geq 0} p_i^{D_i(t)}$$

$$=: \omega(t),$$

where $d(v,t)$ is the number of offspring of vertex v in t, and $D_i(t)$ is the number of vertices in t with i offspring. Thus, $\mathbb{P}\{T_n = t\}$ is proportional to $\omega(t)$.

Example 2.6. If the offspring distribution $(p_i)_{i \in \mathbb{N}_0}$ is the *geometric* distribution $p_i = 2^{-(i+1)}$, $i \in \mathbb{N}_0$, then T_n is uniformly distributed (on the set of rooted ordered trees with n vertices).

Example 2.7. Suppose that the offspring distribution $(p_i)_{i \in \mathbb{N}_0}$ is the *Poisson* distribution $p_i = \frac{e^{-1}}{i!}$, $i \in \mathbb{N}_0$. If we randomly assign the labels $\{1, 2, \ldots, n\}$ to the vertices of T_n and ignore the ordering, then T_n is a uniformly distributed rooted labeled tree with n vertices.

2.3 Finite Trees and Lattice Paths

Although rooted planar trees are not particularly difficult to visualize, we would like to have a quite concrete way of "representing" or "coordinatizing" the planar trees with n vertices that is amenable to investigating the behavior of a random such trees as the number of vertices becomes large. The following simple observation is the key to the work of Aldous, Le Gall and many others on the connections between the asymptotics of large random trees and models for random *paths*.

Given a rooted planar tree with n vertices, start from the root and traverse the tree as follows. At each step move away from the root along the leftmost edge that has not been walked on yet. If this is not possible then step back along the edge leading toward the root. We obtain a with steps of ± 1 by plotting the height (that is, the distance from the root) at each step. Appending a $+1$ step at the beginning and a -1 step at the end gives a lattice excursion path with $2n$ steps that we call the *Harris path* of the tree, although combinatorialists usually call this object a *Dyck path*. We can reverse this procedure and obtain a rooted planar tree with n vertices from any lattice excursion path with $2n$ steps.

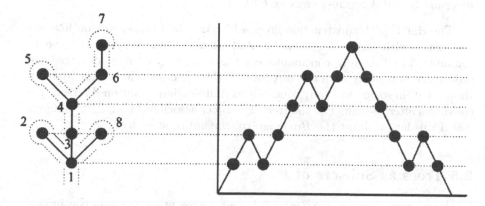

Fig. 2.5. Harris path of a rooted combinatorial tree (figure courtesy of Jim Pitman).

2.4 The Brownian Continuum Random Tree

Put $S_n = \sum_{i=1}^{n} X_i$, where the random variables X_i are independent with $\mathbb{P}\{X_i = \pm 1\} = 1/2$. Conditional on $S_1 = +1$, the path $S_0, S_1, \ldots, S_N$, where $N = \min\{k > 0 : S_k = 0\}$, is the Harris path of the Galton–Watson branching process tree with offspring distribution $p_i = 2^{-(i+1)}$, $i \in \mathbb{N}_0$. Therefore, if we condition on $S_1 = +1$ and $N = 2n$, we get the Harris path of the Galton–Watson branching process tree conditioned on total population size n, and we have observed is the uniform rooted planar tree on n vertices.

We know that suitably re-scaled simple random walk converges to Brownian motion. Similarly, suitably re-scaled simple random walk conditioned to be positive on the first step and return to zero for the first time at time $2n$ converges as $n \to \infty$ to the standard Brownian excursion. Of course, simple random walk is far from being the only process that has Brownian motion as a scaling limit, and so we might hope that there are other random trees with Harris paths that converge to standard Brownian excursion after re-scaling. The following result of Aldous [10] shows that this is certainly the case.

Theorem 2.8. *Let T_n be a conditioned Galton–Watson tree, with offspring mean 1 and variance $0 < \sigma^2 < \infty$. Write $H_n(k)$, $0 \leqslant k \leqslant 2n$ for the Harris path associated with T_n, and interpolate H_n linearly to get a continuous process*

real-valued indexed by the interval $[0, 2n]$ *(which we continue to denote by* H_n*).*
Then, as $n \to \infty$ *through possible sizes of the unconditioned Galton–Watson*
tree,

$$\left(\sigma \frac{H_n(2nu)}{\sqrt{n}}\right)_{0 \leqslant u \leqslant 1} \Rightarrow (2B_u^{ex})_{0 \leqslant u \leqslant 1},$$

where B^{ex} *is the standard Brownian excursion and* $\Rightarrow$ *is the usual weak convergence of probability measures on* $C[0, 1]$.

The Harris path construction gives a bijection between excursion-like lattice paths with steps of ± 1 and rooted planar trees. We will observe in Example 3.14 that any continuous excursion path gives rise to a tree-like object via an analogy with one direction of this bijection. Hence Theorem 2.8 shows that, in some sense, any conditioned finite-variance Galton–Watson tree converges after re-scaling to the tree-like object associated with $2B^{ex}$. Aldous called this latter object the *Brownian continuum random tree* .

2.5 Trees as Subsets of ℓ^1

We have seen in Sections 2.3 and 2.4 that representing trees as continuous paths allows us to use the metric structure on path space to make sense of the idea of a family of random trees converging to some limit random object. In this section we introduce an alternative "coordinatization" of tree-space as the collection of compact subsets of the Banach space $\ell^1 := \{(x_1, x_2, \ldots) : \sum_i |x_i| < \infty\}$. This allows us to use the machinery that has been developed for describing random subsets of a metric space to give another way of expressing such convergence results.

Equip ℓ^1 with the usual norm. Any finite tree with edge lengths can be embedded isometrically as a subset of ℓ^1 (we think of such a tree as a one-dimensional cell complex, that is, as a metric space made up of the vertices of the tree and the connecting edges – not just as the finite metric space consisting of the vertices themselves). For example, the tree of Figure 2.6 is isometric to the set

$$\{te_1 : 0 \leqslant t \leqslant d(\rho, a)\}$$
$$\cup \{d(\rho, d)e_1 + te_2 : 0 \leqslant t \leqslant d(d, b)\}$$
$$\cup \{d(\rho, d)e_1 + d(d, e)e_2 + te_3 : 0 \leqslant t \leqslant d(e, c)\},$$

where $e_1 = (1, 0, 0, \ldots)$, $e_2 = (0, 1, 0, \ldots)$, etc.

Recall Algorithm 2.4 for producing a uniform tree on n labeled vertices. Let $\mathcal{S}^n$ be the subset of ℓ^1 that corresponds to the tree produced by the algorithm. We think of this tree as having edge lengths all equal to 1. More precisely, define a random length random sequence (C_j^n, B_j^n), $0 \leqslant j \leqslant J^n$, as follows:

- $C_0^n = B_0^n := 0$,

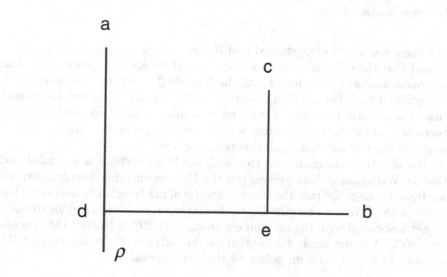

Fig. 2.6. A rooted tree with edge lengths

- C_j^n is the jth element of $\{i : U_i < i - 1\}$,
- $B_j^n := U_{C_j^n}$.

Define $\rho^n : [0, C_{J^n}^n] \to \ell^1$ by $\rho^n(0) := 0$ and

$$\rho^n(x) := \rho^n(B_j^n) + (x - C_j^n)e_{j+1} \text{ for } C_j^n < x \leqslant C_{j+1}^n, \quad 0 \leqslant j \leqslant J^n - 1.$$

Put $\mathcal{S}^n := \rho^n([0, C_{J^n}^n])$.

It is not hard to show that

$$((n^{-1/2}C_1^n, n^{-1/2}B_1^n), (n^{-1/2}C_2^n, n^{-1/2}B_2^n), \ldots)$$
$$\Rightarrow ((C_1, B_1), (C_2, B_2), \ldots),$$

where $\Rightarrow$ denotes weak convergence and $((C_1, B_1), (C_2, B_2), \ldots)$ are defined as follows. Put $C_0 = B_0 := 0$. Let $(C_1, C_2, \ldots)$ be the arrival times in an inhomogeneous Poisson process on $\mathbb{R}_+$ with intensity $t\,dt$. Let $B_j := \xi_j C_j$, where the $\{\xi_j\}_{j \in \mathbb{N}}$ are independent, identically distributed uniform random variables on $[0, 1]$, independent of $\{C_j\}_{j \in \mathbb{N}}$.

Define $\rho : \mathbb{R}_+ \to \ell^1$ by $\rho(0) := 0$ and

$$\rho(x) := \rho(B_j) + (x - C_j)e_{j+1} \text{ for } C_j < x \leqslant C_{j+1}.$$

Set
$$\mathcal{S} := \overline{\bigcup_{t \geqslant 0} \rho([0, t])}.$$

It seems reasonable that
$$n^{-1/2} \mathcal{S}^n \Rightarrow \mathcal{S}$$

in some sense. Aldous [12] showed that $\mathcal{S}$ is almost surely a compact subset of ℓ^1 and that there is convergence in the sense of weak convergence of random compact subsets of ℓ^1 equipped with the *Hausdorff metric* that we will discuss in Section 4.1. Aldous studied $\mathcal{S}$ further in [13, 10]. In particular, he showed that $\mathcal{S}$ is tree-like in various senses: for example, for any two points $x, y \in \mathcal{S}$ there is a unique path connecting x and y (that is, a unique homeomorphic image of the unit interval), and this path has length $\|x - y\|_1$.

Because the uniform rooted tree with n labeled vertices is a conditioned Galton–Watson branching process (for the Poisson offspring distribution), we see from Theorem 2.8 that the Poisson line-breaking "tree" $\mathcal{S}$ is essentially the same as the Brownian continuum random tree, that is, the random tree-like object associated with the random excursion path $2B^{ex}$. In fact, the random tree $\rho(C_n)$ has the same distribution as the subtree of the Brownian CRT spanned by n i.i.d. uniform points on the unit interval.

3

ℝ-Trees and 0-Hyperbolic Spaces

3.1 Geodesic and Geodesically Linear Metric Spaces

We follow closely the development in [39] in this section and leave some of the more straightforward proofs to the reader.

Definition 3.1. *A* segment *in a metric space (X, d) is the image of an isometry $\alpha : [a, b] \to X$. The end points of the segment are $\alpha(a)$ and $\alpha(b)$.*

Definition 3.2. *A metric space (X, d) is* geodesic *if for all $x, y \in X$, there is a segment in X with endpoints $\{x, y\}$, and (X, d) is* geodesically linear *if, for all $x, y \in X$, there is a **unique** segment in X with endpoints $\{x, y\}$.*

Example 3.3. Euclidean space $\mathbb{R}^d$ is geodesically linear. The closed annulus $\{z \in \mathbb{R}^2 : 1 \leqslant |z| \leqslant 2\}$ is not geodesic in the metric inherited from $\mathbb{R}^2$, but it is geodesic in the metric defined by taking the infimum of the Euclidean lengths of piecewise-linear paths between two points. The closed annulus is not geodesically linear in this latter metric: for example, a pair of points of the form z and $-z$ are the endpoints of two segments – see Figure 3.1. The open annulus $\{z \in \mathbb{R}^2 : 1 < |z| < 2\}$ is not geodesic in the metric defined by taking the infimum over all piecewise-linear paths between two points: for example, there is no segment that has a pair of points of the form z and $-z$ as its endpoints.

Lemma 3.4. *Consider a metric space (X, d). Let σ be a segment in X with endpoints x and z, and let τ be a segment in X with endpoints y and z.*

(a) Suppose that $d(u, v) = d(u, z) + d(z, v)$ for all $u \in \sigma$ and $v \in \tau$. Then $\sigma \cup \tau$ is a segment with endpoints x and y.

(b) Suppose that $\sigma \cap \tau = \{z\}$ and $\sigma \cup \tau$ is a segment. Then $\sigma \cup \tau$ has endpoints x and y.

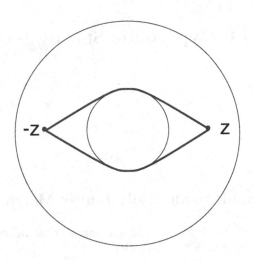

Fig. 3.1. Two geodesics with the same endpoints in the intrinsic path length metric on the annulus

Lemma 3.5. *Let (X, d) be a geodesic metric space such that if two segments of (X, d) intersect in a single point, which is an endpoint of both, then their union is a segment. Then (X, d) is a geodesically linear metric space.*

Proof. Let σ, τ be segments, both with endpoints u, v. Fix $w \in \sigma$, and define w' to be the point of τ such that $d(u, w) = d(u, w')$ (so that $d(v, w) = d(v, w')$). We have to show $w = w'$.

Let ρ be a segment with endpoints w, w'. Now $\sigma = \sigma_1 \cup \sigma_2$, where σ_1 is a segment with endpoints u, w, and σ_2 is a segment with endpoints w, v – see Figure 3.2.

We claim that either $\sigma_1 \cap \rho = \{w\}$ or $\sigma_2 \cap \rho = \{w\}$. This is so because if $x \in \sigma_1 \cap \rho$ and $y \in \sigma_2 \cap \rho$, then $d(x, y) = d(x, w) + d(w, y)$, and either $d(w, y) = d(w, x) + d(x, y)$ or $d(w, x) = d(w, y) + d(x, y)$, depending on how x, y are situated in the segment ρ. It follows that either $x = w$ or $y = w$, establishing the claim.

Now, if $\sigma_1 \cap \rho = \{w\}$, then, by assumption, $\sigma_1 \cup \rho$ is a segment, and by Lemma 3.4(b) its endpoints are u, w'. Since $w \in \sigma_1 \cup \rho$, $d(u, w') = d(u, w) + d(w, w')$, so $w = w'$. Similarly, if $\sigma_2 \cap \rho = \{w\}$ then $w = w'$.   □

Lemma 3.6. *Consider a geodesically linear metric space (X, d).*

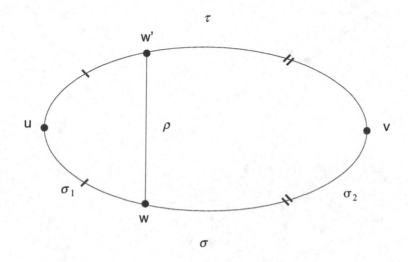

Fig. 3.2. Construction in the proof of Lemma 3.5

(i)   *Given points $x, y, z \in X$, write σ for the segment with endpoints x, y.*
      Then $z \in \sigma$ if and only if $d(x, y) = d(x, z) + d(z, y)$.
(ii) *The intersection of two segments in X is also a segment if it is non-*
      empty.
(iii) *Given $x, y \in X$, there is a unique isometry $\alpha : [0, d(x, y)] \to X$ such that*
      $\alpha(0) = x$ and $\alpha(d(x, y)) = y$. Write $[x, y]$ for the resulting segment. If
      $u, v \in [x, y]$, then $[u, v] \subseteq [x, y]$.

3.2 0-Hyperbolic Spaces

Definition 3.7. *For x, y, v in a metric space (X, d), set*

$$(x \cdot y)_v := \frac{1}{2}(d(x, v) + d(y, v) - d(x, y))$$

– see Figure 3.3.

Remark 3.8. For $x, y, v, t \in X$,

$$0 \leqslant (x \cdot y)_v \leqslant d(x, v) \wedge d(y, v)$$

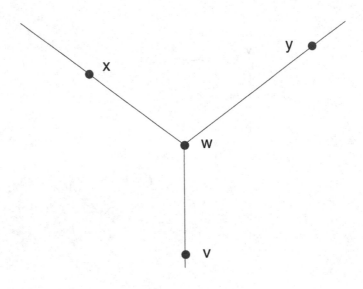

Fig. 3.3. $(x \cdot y)_v = d(w, v)$ in this tree

and
$$(x \cdot y)_t = d(t, v) + (x \cdot y)_v - (x \cdot t)_v - (y \cdot t)_t.$$

Definition 3.9. *A metric space* (X, d) *is* 0-hyperbolic with respect to v *if for all* $x, y, z \in X$
$$(x \cdot y)_v \geqslant (x \cdot z)_v \wedge (y \cdot z)_v$$
– *see Figure 3.4.*

Lemma 3.10. *If the metric space* (X, d) *is 0-hyperbolic with respect to some point of* X, *then* (X, d) *is 0-hyperbolic with respect to all points of* X.

Remark 3.11. In light of Lemma 3.10, we will refer to a metric space that is 0-hyperbolic with respect to one, and hence all, of its points as simply being 0-hyperbolic. Note that any subspace of a 0-hyperbolic metric space is also 0-hyperbolic.

Lemma 3.12. *The metric space* (X, d) *is 0-hyperbolic if and only if*
$$d(x, y) + d(z, t) \leqslant \max\{d(x, z) + d(y, t), \ d(y, z) + d(x, t)\}$$
for all $x, y, z, t \in X$,

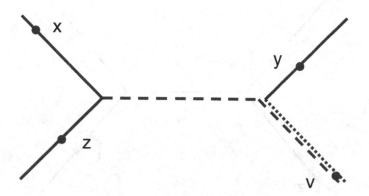

Fig. 3.4. The 0-hyperbolicity condition holds for this tree. Here $(x \cdot y)_v$ and $(y \cdot z)_v$ are both given by the length of the dotted segment, and $(x \cdot z)_v$ is the length of the dashed segment. Note that $(x \cdot y)_v \geq (x \cdot z)_v \wedge (y \cdot z)_v$, with similar inequalities when x, y, z are permuted.

Remark 3.13. The set of inequalities in Lemma 3.12 is usually called the *four-point condition* – see Figure 3.5.

Example 3.14. Write $C(\mathbb{R}_+)$ for the space of continuous functions from $\mathbb{R}_+$ into $\mathbb{R}$. For $e \in C(\mathbb{R}_+)$, put $\zeta(e) := \inf\{t > 0 : e(t) = 0\}$ and write

$$U := \left\{ e \in C(\mathbb{R}_+) : \begin{array}{l} e(0) = 0, \ \zeta(e) < \infty, \\ e(t) > 0 \text{ for } 0 < t < \zeta(e), \\ \text{and } e(t) = 0 \text{ for } t \geq \zeta(e) \end{array} \right\}$$

for the space of positive excursion paths. Set $U^\ell := \{e \in U : \zeta(e) = \ell\}$.

We associate each $e \in U^\ell$ with a compact metric space as follows. Define an equivalence relation $\sim_e$ on $[0, \ell]$ by letting

$$u_1 \sim_e u_2, \quad \text{iff} \quad e(u_1) = \inf_{u \in [u_1 \wedge u_2, u_1 \vee u_2]} e(u) = e(u_2).$$

Consider the following semi-metric on $[0, \ell]$

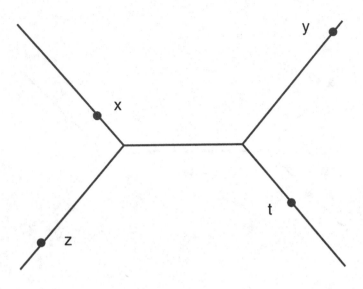

Fig. 3.5. The four-point condition holds on a tree: $d(x,z) + d(y,t) \leqslant d(x,y) + d(z,t) = d(x,t) + d(y,z)$

$$d_{T_e}(u_1, u_2) := e(u_1) - 2 \inf_{u \in [u_1 \wedge u_2, u_1 \vee u_2]} e(u) + e(u_2),$$

that becomes a true metric on the quotient space $T_e := [0, \ell]\big|_{\sim_e}$ – see Figure 3.6.

It is straightforward to check that the quotient map from $[0, \ell]$ onto T_e is continuous with respect to d_{T_e}. Thus, (T_e, d_{T_e}) is path-connected and compact as the continuous image of a metric space with these properties. In particular, (T_e, d_{T_e}) is complete. It is not difficult to check that (T_e, d_{T_e}) satisfies the four-point condition, and, hence, is 0-hyperbolic.

3.3 ℝ-Trees

3.3.1 Definition, Examples, and Elementary Properties

Definition 3.15. *An* ℝ-tree *is a metric space* (X, d) *with the following properties.*

Axiom (a) The space (X, d) *is geodesic.*

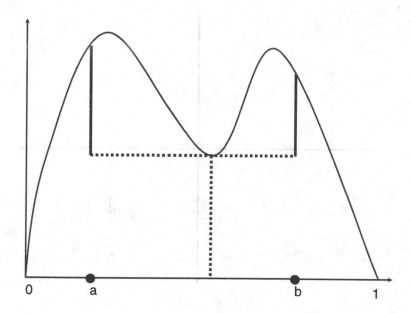

Fig. 3.6. An excursion path on $[0, 1]$ determines a distance between the points a and b

Axiom (b) If two segments of (X, d) intersect in a single point, which is an endpoint of both, then their union is a segment.

Example 3.16. Finite trees with edge lengths (sometimes called weighted trees) are examples of ℝ-trees. To be a little more precise, we don't think of such a tree as just being its finite set of vertices with a collection of distances between them, but regard the edges connecting the vertices as also being part of the metric space.

Example 3.17. Take X to be the plane $\mathbb{R}^2$ equipped with the metric

$$d((x_1, x_2), (y_1, y_2)) := \begin{cases} |x_2 - y_2|, & \text{if } x_1 = y_1, \\ |x_1 - y_1| + |x_2| + |y_2|, & \text{if } x_1 \neq y_1. \end{cases}$$

That is, we think of the plane as being something like the skeleton of a fish, in which the horizontal axis is the spine and vertical lines are the ribs. In order to compute the distance between two points on different ribs, we use the length of the path that goes from the first point to the spine, then along the spine to the rib containing the second point, and then along that second rib – see Figure 3.7.

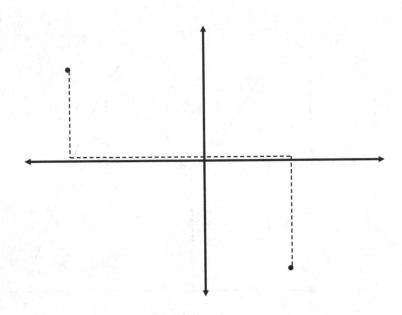

Fig. 3.7. The distance between two points of $\mathbb{R}^2$ in the metric of Example 3.17 is the (Euclidean) length of the dashed path

Example 3.18. Consider the collection $\mathcal{T}$ of bounded subsets of $\mathbb{R}$ that contain their supremum. We can think of the elements of $\mathcal{T}$ as being arrayed in a tree–like structure in the following way. Using genealogical terminology, write $h(B) := \sup B$ for the real–valued *generation* to which $B \in \mathcal{T}$ belongs and $B|t := (B \cap \,]-\infty, t]) \cup \{t\} \in \mathcal{T}$ for $t \leqslant h(B)$ for the *ancestor* of B in generation t. For $A, B \in \mathcal{T}$ the generation of the *most recent common ancestor* of A and B is $\tau(A, B) := \sup\{t \leqslant h(A) \wedge h(B) : A|t = B|t\}$. That is, $\tau(A, B)$ is the generation at which the lineages of A and B diverge. There is a natural genealogical distance on $\mathcal{T}$ given by

$$D(A, B) := [h(A) - \tau(A, B)] + [h(B) - \tau(A, B)].$$

See Figure 3.8.

It is not difficult to show that the metric space $(\mathcal{T}, D)$ is a ℝ-tree. For example, the segment with end-points A and B is the set $\{A|t : \tau(A, B) \leqslant t \leqslant h(A)\} \cup \{B|t : \tau(A, B) \leqslant t \leqslant h(B)\}$.

The metric space $(\mathcal{T}, D)$ is essentially "the" real tree of [47, 137] (the latter space has as its points the bounded subsets of $\mathbb{R}$ that contain their infimum and the corresponding metric is such that the map from $(\mathcal{T}, D)$ into this latter space given by $A \mapsto -A$ is an isometry). With a slight abuse of

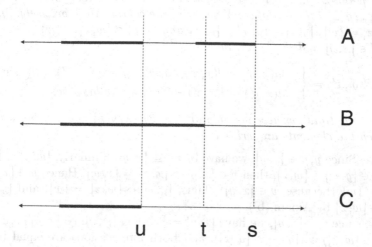

Fig. 3.8. The set C is the most recent common ancestor of the sets $A, B \subset \mathbb{R}$ thought of as points of "the" real tree of Example 3.18. The distance $D(A, B)$ is $[s - u] + [t - u]$.

nomenclature, we will refer here to $(\mathcal{T}, D)$ as the real tree. Note that $(\mathcal{T}, D)$ is huge: for example, the removal of any point shatters $\mathcal{T}$ into uncountably many connected components.

Example 3.19. We will see in Example 3.37 that the compact 0-hyperbolic metric space (T_e, d_{T_e}) of Example 3.14 that arises from an excursion path $e \in U$ is a ℝ-tree.

The following result is a consequence of Axioms (a) and (b) and Lemma 3.5.

Lemma 3.20. *An ℝ-tree is geodesically linear. Moreover, if (X, d) is a ℝ-tree and $x, y, z \in X$ then $[x, y] \cap [x, z] = [x, w]$ for some unique $w \in X$.*

Remark 3.21. It follows from Lemma 3.4, Lemma 3.6 and Lemma 3.20 that Axioms (a) and (b) together imply following condition that is stronger than Axiom (b):

Axiom (b') If (X, d) is a ℝ-tree, $x, y, z \in X$ and $[x, y] \cap [x, z] = \{x\}$, then
$[x, y] \cup [x, z] = [y, z]$

Lemma 3.22. *Let x, y, z be points of a ℝ-tree (X, d), and write w for the unique point such that $[x, y] \cap [x, z] = [x, w]$.*

(i) The points x, y, z, w and the segments connecting them form a Y shape, with x, y, z at the tips of the Y and w at the center. More precisely, $[y, w] \cap [w, z] = \{w\}$, $[y, z] = [y, w] \cup [w, z]$ and $[x, y] \cap [w, z] = \{w\}$.

(ii) If $y' \in [x, y]$ and $z' \in [x, z]$, then

$$d(y', z') = \begin{cases} |d(x, y') - d(x, z')|, & \text{if } d(x, y') \wedge d(x, z') \leqslant d(x, w), \\ d(x, y') + d(x, z') - 2d(x, w), & \text{otherwise.} \end{cases}$$

(iii) The "centroid" w depends only on the set $\{x, y, z\}$, not on the order in which the elements are written.

Proof. (i) Since $y, w \in [x, y]$, we have $[y, w] \subseteq [x, y]$. Similarly, $[w, z] \subseteq [x, z]$. So, if $u \in [y, w] \cap [w, z]$, then $u \in [x, y] \cap [x, z] = [x, w]$. Hence $u \in [x, w] \cap [y, w] = \{w\}$ (because $w \in [x, y]$). Thus, $[y, w] \cap [w, z] = \{w\}$, and $[y, z] = [y, w] \cup [w, z]$ by Axiom (b').

Now, since $w \in [x, y]$, we have $[x, y] = [x, w] \cup [w, y]$, so $[x, y] \cap [w, z] = ([x, w] \cap [w, z]) \cup ([y, w] \cap [w, z])$, and both intersections are equal to $\{w\}$ ($w \in [x, z]$).

(ii) If $d(x, y') \leqslant d(x, w)$ then $y', z' \in [x, z]$, and so $d(y', z') = |d(x, y') - d(x, z')|$. Similarly, if $d(x, z') \leqslant d(x, w)$, then $y', z' \in [x, y]$, and once again $d(y', z') = |d(x, y') - d(x, z')|$.

If $d(x, y') > d(x, w)$ and $d(x, z') > d(x, w)$, then $y' \in [y, w]$ and $z' \in [z, w]$. Hence, by part (i),

$$\begin{aligned} d(y', z') &= d(y', w) + d(w, z') \\ &= (d(x, y') - d(x, w)) + (d(x, z') - d(x, w) \\ &= d(x, y') + d(x, z') - 2d(x, w). \end{aligned}$$

(iii) We have by part (i) that

$$\begin{aligned} [y, x] \cap [y, z] &= [y, x] \cap ([y, w] \cup [w, z]) \\ &= [y, w] \cup ([y, x] \cap [w, z]) \\ &= [y, w] \cup ([y, w] \cap [w, z]) \cup ([w, x] \cap [w, z]) \end{aligned}$$

Now $[y, w] \cap [w, z] = \{w\}$ by part (1) and $[w, x] \cap [w, z] = \{w\}$ since $w \in [x, z]$. Hence, $[y, x] \cap [y, z] = [y, w]$. Similarly, $[z, x] \cap [z, y] = [z, w]$, and part (iii) follows. □

Definition 3.23. *In the notation of Lemma 3.22, write $Y(x, y, z) := w$ for the centroid of $\{x, y, z\}$.*

Remark 3.24. Note that we have

$$[x, y] \cap [w, z] = [x, z] \cap [w, y] = [y, z] \cap [w, x] = \{w\}.$$

Also, $d(x, w) = (y \cdot z)_x$, $d(y, w) = (x \cdot z)_y$, and $d(z, w) = (x \cdot y)_z$. In Figure 3.3, $Y(x, y, v) = w$.

Corollary 3.25. *Consider a ℝ-tree (X, d) and points $x_0, x_1, \ldots, x_n \in X$. The segment $[x_0, x_n]$ is a subset of $\bigcup_{i=1}^n [x_{i-1}, x_i]$.*

Proof. If $n = 2$, then, by Lemma 3.22,

$$[x_0, x_2] = [x_0, Y(x_0, x_1, x_2)] \cup [Y(x_0, x_1, x_2), x_2] \subseteq [x_0, x_1] \cup [x_1, x_2].$$

If $n > 2$, then $[x_0, x_n] \subseteq [x_0, x_{n-1}] \cup [x_{n-1}, x_n]$ by the case $n = 2$, and the result follows by induction on n.   □

Lemma 3.26. *Consider a ℝ-tree (X, d). Let $\alpha : [a, b] \to X$ be a continuous map. If $x = \alpha(a)$ and $y = \alpha(b)$, then $[x, y]$ is a subset of the image of α.*

Proof. Let A denote the image of α. Since A is a closed subset of X (being compact as the image of a compact interval by a continuous map), it is enough to show that every point of $[x, y]$ is within distance ϵ of A, for all $\epsilon > 0$.

Given $\epsilon > 0$, the collection $\{\alpha^{-1}(B(x, \epsilon/2)) : x \in A\}$ is an open covering of the compact metric space $[a, b]$, so there is a number $\delta > 0$ such that any two points of $[a, b]$ that are distance less than δ apart belong to some common set in the cover.

Choose a partition of $[a, b]$, say $a = t_0 < \cdots < t_n = b$, so that for $1 \leqslant i \leqslant n$ we have $t_i - t_{i-1} < \delta$, and, therefore, $d(\alpha(t_{i-1}), \alpha(t_i)) < \epsilon$. Then all points of $[\alpha(t_{i-1}), \alpha(t_i)]$ are at distance less than ϵ from $\{\alpha(t_{i-1}), \alpha(t_i)\} \subseteq A$ for $1 \leqslant i \leqslant n$. Finally, $[x, y] \subseteq \bigcup_{i=1}^n [\alpha(t_{i-1}), \alpha(t_i)]$, by Corollary 3.25.   □

Definition 3.27. *For points $x_0, x_1, \ldots, x_n$ in a ℝ-tree (X, d), write $[x_0, x_n] = [x_0, x_1, \ldots, x_n]$ to mean that, if $\alpha : [0, d(x_0, x_n)] \to X$ is the unique isometry with $\alpha(0) = x_0$ and $\alpha(d(x_0, x_n)) = x_n$, then $x_i = \alpha(a_i)$, for some $a_0, a_1, a_2, \ldots, a_n$ with $0 = a_0 \leqslant a_1 \leqslant a_2 \leqslant \cdots \leqslant a_n = d(x_0, x_n)$.*

Lemma 3.28. *Consider a ℝ-tree (X, d). If $x_0, \ldots, x_n \in X$, $x_i \neq x_{i+1}$ for $1 \leqslant i \leqslant n - 2$ and $[x_{i-1}, x_i] \cap [x_i, x_{i+1}] = \{x_i\}$ for $1 \leqslant i \leqslant n - 1$, then $[x_0, x_n] = [x_0, x_1, \ldots, x_n]$.*

Proof. There is nothing to prove if $n \leqslant 2$. Suppose $n = 3$. We can assume $x_0 \neq x_1$ and $x_2 \neq x_3$, otherwise there is again nothing to prove. Let $w = Y(x_0, x_2, x_3)$.

Now $w \in [x_0, x_2]$ and $x_1 \in [x_0, x_2]$, so $[x_2, w] \cap [x_2, x_1] = [x_2, v]$, where v is either w or x_1, depending on which is closer to x_2. But $[x_2, w] \cap [x_2, x_1] \subseteq [x_2, x_3] \cap [x_2, x_1] = \{x_2\}$, so $v = x_2$.

Since $x_1 \neq x_2$, we conclude that $w = x_2$. Hence $[x_0, x_2] \cap [x_2, x_3] = \{x_2\}$, which implies $[x_0, x_3] = [x_0, x_2, x_3] = [x_0, x_1, x_2, x_3]$.

Now suppose $n > 3$. By induction,

$$[x_0, x_{n-1}] = [x_0, x_1, \ldots, x_{n-2}, x_{n-1}] = [x_0, x_{n-2}, x_{n-1}].$$

By the $n = 3$ case,

$$[x_0, x_n] = [x_0, x_{n-2}, x_{n-1}, x_n] = [x_0, x_1, \ldots, x_{n-2}, x_{n-1}, x_n]$$

as required.   □

3.3.2 ℝ-Trees are 0-Hyperbolic

Lemma 3.29. *A ℝ-tree (X, d) is 0-hyperbolic.*

Proof. Fix $v \in X$. We have to show

$$(x \cdot y)_v \geqslant (x \cdot z)_v \wedge (y \cdot z)_v$$
$$(x \cdot z)_v \geqslant (x \cdot y)_v \wedge (y \cdot z)_v$$
$$(y \cdot z)_v \geqslant (x \cdot y)_v \wedge (x \cdot z)_v$$

for all x, y, z. Note that if this is so, then one of $(x \cdot y)_v, (x \cdot z)_v, (y \cdot z)_v$ is at least as great as the other two, which are equal.

Let $q = Y(x, v, y)$, $r = Y(y, v, z)$, and $s = Y(z, v, x)$. We have $(x \cdot y)_v = d(v, q)$, $(y \cdot z)_v = d(v, s)$, and $(z \cdot x)_v = d(v, r)$. We may assume without loss of generality that

$$d(v, q) \leqslant d(v, r) \leqslant d(v, s),$$

in which case have to show that $q = r$ – see Figure 3.9.

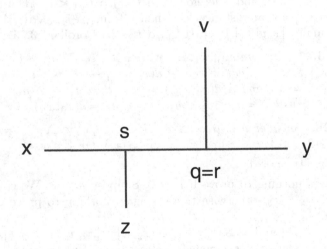

Fig. 3.9. The configuration demonstrated in the proof of Lemma 3.29

Now $r, s \in [v, z]$ by definition, and $d(v, r) \leqslant d(v, s)$, so that $[v, s] = [v, r, s]$. Also, by definition of s, $[v, x] = [v, s, x] = [v, r, s, x]$. Hence $r \in [v, x] \cap [v, y] = [v, q]$. Since $d(v, q) \leqslant d(v, r)$, we have $q = r$, as required.   □

Remark 3.30. Because any subspace of a 0-hyperbolic space is still 0-hyperbolic, we can't expect that the converse to Lemma 3.29 holds. However, we will see in Theorem 3.38 that any 0-hyperbolic space is isometric to a subspace of a ℝ-tree.

3.3.3 Centroids in a 0-Hyperbolic Space

Definition 3.31. *A set* $\{a, b, c\} \subset \mathbb{R}$ *is called an* isosceles triple *if*

$$a \geqslant b \wedge c, \ b \geqslant c \wedge a, \ \text{and} \ c \geqslant a \wedge b.$$

(This means that at least two of a, b, c are equal, and not greater than the third.)

Remark 3.32. The metric space (X, d) is 0-hyperbolic if and only if $(x \cdot y)_v, (x \cdot z)_v, (y \cdot z)_v$ is an isosceles triple for all $x, y, z, v \in X$.

Lemma 3.33. *(i) If $\{a, b, c\}$ is any triple then*

$$\{a \wedge b, b \wedge c, c \wedge a\}$$

is an isosceles triple.
(ii) If $\{a, b, c\}$ and $\{d, e, f\}$ are isosceles triples then so is

$$\{a \wedge d, b \wedge e, c \wedge f\}.$$

Lemma 3.34. *Consider a 0-hyperbolic metric space (X, d). Let σ, τ be segments in X with endpoints v, x and v, y respectively. Write $x \cdot y := (x \cdot y)_v$.*

(i) If $x' \in \sigma$, then $x' \in \tau$ if and only if $d(v, x') \leqslant x \cdot y$.
(ii) If w is the point of σ at distance $x \cdot y$ from v, then $\sigma \cap \tau$ is a segment with endpoints v and w.

Proof. If $d(x', v) > d(y, v)$ then $x' \notin \tau$, and $d(x', v) > x \cdot y$, so we can assume that $d(x', v) < d(y, v)$. Let y' be the point in τ such that $d(v, x') = d(v, y')$. Define

$$\alpha = x \cdot y, \ \beta = x' \cdot y, \ \gamma = x \cdot x', \ \alpha' = x' \cdot y'.$$

Since $x' \in \sigma$ and $y' \in \tau$, we have $\gamma = d(v, x') = d(v, y') = y \cdot y'$. Hence, (α, β, γ) and (α', β, γ) are isosceles triples. We have to show that $x' \in \tau$ if and only if $\alpha \geqslant \gamma$. The two cases $\alpha < \gamma$ and $\alpha \geqslant \gamma$ are illustrated in Figure 3.10 and Figure 3.11 respectively.

Now,

$$\beta = x' \cdot y \leqslant d(v, x') = x \cdot x' = \gamma.$$

Also,

$$\alpha' = d(v, x') - \frac{1}{2}d(x', y') = \gamma - \frac{1}{2}d(x', y') \leqslant \gamma$$

and

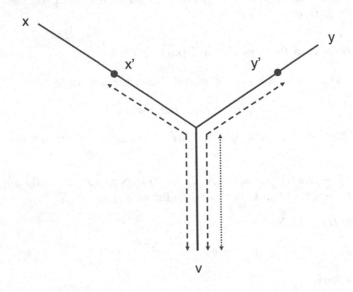

Fig. 3.10. First case of the construction in the proof of Lemma 3.34. Here γ is either of the two equal dashed lengths and $\alpha = \beta = \alpha'$ is the dotted length. As claimed, $\alpha < \gamma$ and $x' \notin \tau$.

$$x' \in \tau \Leftrightarrow x' = y' \Leftrightarrow d(x', y') = 0 \Leftrightarrow \alpha' = \gamma.$$

Moreover, $\alpha' = \gamma$ if and only if $\beta = \gamma$, because (α', β, γ) is an isosceles triple and $\alpha', \beta \leqslant \gamma$. Since (α, β, γ) is also an isosceles triple, the equality $\beta = \gamma$ is equivalent to the inequality $\alpha \geqslant \gamma$. This proves part (i). Part (ii) of the lemma follows immediately. $\qquad\square$

Lemma 3.35. *Consider a 0-hyperbolic metric space* (X, d). *Let* σ, τ *be segments in* X *with endpoints* v, x *and* v, y *respectively. Set* $x \cdot y := (x \cdot y)_v$. *Write* w *for the point of* σ *at distance* $x \cdot y$ *from* v *(so that* w *is an endpoint of* $\sigma \cap \tau$ *by Lemma 3.34). Consider two points* $x' \in \sigma$, $y' \in \tau$, *and suppose* $d(x', v) \geqslant x \cdot y$ *and* $d(y', v) \geqslant x \cdot y$. *Then*

$$d(x', y') = d(x', w) + d(y', w).$$

Proof. The conclusion is clear if $d(x', v) = x \cdot y$ (when $x' = w$) or $d(y', v) = x \cdot y$ (when $y' = w$), so we assume that $d(x', v) > x \cdot y$ and $d(y', v) > x \cdot y$. As in the proof of Lemma 3.34, we put

$$\alpha = x \cdot y, \quad \beta = x' \cdot y, \quad \gamma = x \cdot x', \quad \alpha' = x' \cdot y',$$

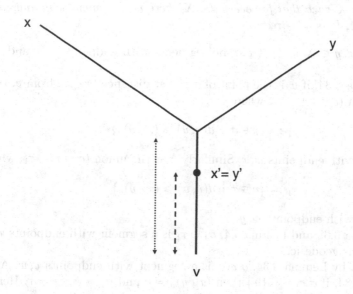

Fig. 3.11. Second case of the construction in the proof of Lemma 3.34. Here $\gamma = \beta = \alpha'$ is the dashed length and α is the dotted length. As claimed, $\alpha \geqslant \gamma$ and $x' \in \tau$.

and we also put $\gamma' = y \cdot y'$, so that $\gamma = d(v, x')$ and $\gamma' = d(v, y')$. Thus, $\alpha < \gamma$. Hence, $\alpha = \beta$ since (α, β, γ) is an isosceles triple. Also, $\alpha < \gamma'$, so that $\beta < \gamma'$. Hence, $\alpha = \alpha' = \beta$ because (α', β, γ) is an isosceles triple.

By definition of α',

$$d(x', y') = d(v, x') + d(v, y') - 2\alpha'$$
$$= d(v, x') + d(v, y') - 2\alpha.$$

Since $w \in \sigma \cap \tau$, $\alpha = d(v, w) < d(v, x'), d(v, y')$ and σ, τ are segments, it follows that

$$d(x', w) = d(v, x') - \alpha$$

and

$$d(y', w) = d(v, y') - \alpha,$$

and the lemma follows on adding these equations.   □

3.3.4 An Alternative Characterization of ℝ-Trees

Lemma 3.36. *Consider a 0-hyperbolic metric space (X, d). Suppose that there is a point $v \in X$ such that for every $x \in X$ there is a segment with endpoints v, x. Then (X, d) is a ℝ-tree.*

Proof. Take $x, y \in X$ and let σ, τ be segments with endpoints v, x and v, y respectively.

By Lemma 3.34, if w is the point of $\sigma \cap \tau$ at distance $(x \cdot y)_v$ from v, then σ is the union $(\sigma \cap \tau) \cup \sigma_1$, where

$$\sigma_1 := \{u \in \sigma : d(v, u) \geqslant (x \cdot y)_v\}$$

is a segment with endpoints w, x. Similarly, τ is the union $(\sigma \cap \tau) \cup \tau_1$, where

$$\tau_1 := \{u \in \tau : d(v, u) \geqslant (x \cdot y)_v\}$$

is a segment with endpoints w, y.

By Lemma 3.35 and Lemma 3.4, $\sigma_1 \cup \tau_1$ is a segment with endpoints x, y. Thus, (X, d) is geodesic.

Note that by Lemma 3.34, $\sigma \cap \tau$ is a segment with endpoints v, w. Also, by Lemma 3.34, if $\sigma \cap \tau = \{w\}$ then $(x \cdot y)_v = 0$ and $\sigma_1 = \sigma$, $\tau_1 = \tau$. Hence, $\sigma \cup \tau$ is a segment. Now, by Lemma 3.10, we may replace v in this argument by any other point of X. Hence, (X, d) satisfies the axioms for a ℝ-tree. □

Example 3.37. We noted in Example 3.14 that the compact metric space (T_e, d_{T_e}) that arises from an excursion path $e \in U$ is 0-hyperbolic. We can use Lemma 3.36 to show that (T_e, d_{T_e}) is a ℝ-tree. Suppose that $e \in U^\ell$. Take $x \in T_e$ and write t for a point in $[0, \ell]$ such that x is the image of t under the quotient map from $[0, \ell]$ onto T_e. Write $v \in T_e$ for the image of $0 \in [0, \ell]$ under the quotient map from $[0, \ell]$ onto T_e. Note that v is also the image of $\ell \in [0, \ell]$. For $h \in [0, e(t)]$, set $\lambda_h := \sup\{s \in [0, t] : e(s) = h\}$. Then the image of the set $\{\lambda_h : h \in [0, e(t)]\} \subseteq [0, \ell]$ under the quotient map is a segment in T_e that has endpoints v and x.

3.3.5 Embedding 0-Hyperbolic Spaces in ℝ-Trees

Theorem 3.38. *Let (X, d) be a 0-hyperbolic metric space. There exists a ℝ-tree (X', d') and an isometry $\phi : X \to X'$.*

Proof. Fix $v \in X$. Write $x \cdot y := (x \cdot y)_v$ for $x, y \in X$. Let

$$Y = \{(x, m) : x \in X, m \in \mathbb{R} \text{ and } 0 \leqslant m \leqslant d(v, x)\}.$$

Define, for $(x, m), (y, n) \in Y$,

$$(x, m) \sim (y, n) \text{ if and only if } x \cdot y \geqslant m = n.$$

This is an equivalence relation on Y. Let $X' = Y/\sim$, and let $\langle x, m \rangle$ denote the equivalence class of (x, m). We define the metric by

$$d'(\langle x, m \rangle, \langle y, n \rangle) = m + n - 2[m \wedge n \wedge (x \cdot y)].$$

The construction is illustrated in Figure 3.12.

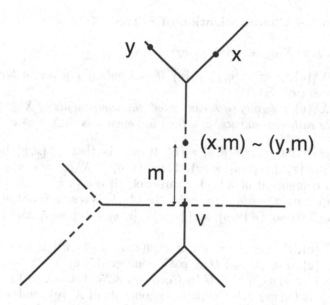

Fig. 3.12. The embedding of Theorem 3.38. Solid lines represent points that are in X, while dashed lines represent points that are added to form X'.

It follows by assumption that d' is well defined. Note that

$$d'(\langle x, m \rangle, \langle x, n \rangle) = |m - n|$$

and $\langle x, 0 \rangle = \langle v, 0 \rangle$ for all $x \in X$, so $d'(\langle x, m \rangle, \langle v, 0 \rangle) = m$. Clearly d' is symmetric, and it is easy to see that $d'(\langle x, m \rangle, \langle y, n \rangle) = 0$ if and only if $\langle x, m \rangle = \langle y, n \rangle$. Also, in X',

$$(\langle x, m \rangle \cdot \langle y, n \rangle)_{\langle v, 0 \rangle} = m \wedge n \wedge (x \cdot y).$$

If $\langle x, m \rangle, \langle y, n \rangle$ and $\langle z, p \rangle$ are three points of X', then

$$\{m \wedge n, n \wedge p, p \wedge m\}$$

is an isosceles triple by Lemma 3.33(1). Hence, by Lemma 3.33(2), so is $\{m \wedge n \wedge (x \cdot y), n \wedge p \wedge (y \cdot z), p \wedge m \wedge (z \cdot x)\}$. It follows that (X', d') is a 0-hyperbolic metric space.

If $\langle x, m \rangle \in X'$, then the mapping $\alpha : [0, m] \to X'$ given by $\alpha(n) = \langle x, n \rangle$ is an isometry, so the image of α is a segment with endpoints $\langle v, 0 \rangle$ and $\langle x, m \rangle$. It now follows from Lemma 3.36 that (X', d') is a ℝ-tree. Further, the mapping $\phi : X \to X'$ defined by $\phi(x) = \langle x, d(v, x) \rangle$ is easily seen to be an isometry.   □

3.3.6 Yet another Characterization of ℝ-Trees

Lemma 3.39. *Let (X, d) be a ℝ-tree. Fix $v \in X$.*

(i) For $x, y \in X \backslash \{v\}$, $[v, x] \cap [v, y] \neq \{v\}$ if and only if x, y are in the same path component of $X \backslash \{v\}$.

(ii) The space $X \backslash \{v\}$ is locally path connected, the components of $X \backslash \{v\}$ coincide with its path components, and they are open sets in X.

Proof. (i) Suppose that $[v, x] \cap [v, y] \neq \{v\}$. It can't be that $v \in [x, y]$, because that would imply $[x, v] \cap [v, y] = \{v\}$. Thus, $[x, y] \subseteq X \backslash \{v\}$ and x, y are in the same path component of $X \backslash \{v\}$. Conversely, if $\alpha : [a, b] \to X \backslash \{v\}$ is a continuous map, with $x = \alpha(a)$, $y = \alpha(b)$, then $[a, b]$ is a subset of the image of α by Lemma 3.26, so $v \notin [x, y]$, and $[v, x] \cap [v, y] \neq \{v\}$ by Axiom (b') for a ℝ-tree.

(ii) For $x \in X \backslash \{v\}$, the set $U := \{y \in X : d(x, y) < d(x, v)\}$ is an open set in X, $U \subseteq X \backslash \{v\}$, $x \in U$, and U is path connected. For if $y, z \in U$, then $[x, y] \cup [x, z] \subseteq U$, and so $[y, z] \subseteq U$ by Corollary 3.25. Thus, $X \backslash \{v\}$ is locally path connected. It follows that the path components of $X \backslash \{v\}$ are both open and closed, and (ii) follows easily.   □

Theorem 3.40. *A metric space (X, d) is a ℝ-tree if and only if it is connected and 0-hyperbolic.*

Proof. An ℝ-tree is geodesic, so it is path connected. Hence, it is connected. Therefore, it is 0-hyperbolic by Lemma 3.29.

Conversely, assume that a metric space (X, d) is connected and 0-hyperbolic. By Theorem 3.38 there is an embedding of (X, d) in a ℝ-tree (X', d'). Let $x, y \in X$, suppose $v \in X' \backslash X$ and $v \in [x, y]$. Then $[v, x] \cap [v, y] = \{v\}$ and so by Lemma 3.39, x, y are in different components of $X \backslash \{v\}$.

Let C be the component of $X \backslash \{v\}$ containing x. By Lemma 3.39, C is open and closed, so $X \cap C$ is open and closed in X. Since $x \in X \cap C$, $y \notin X \cap C$, this contradicts the connectedness of X. Thus, $[x, y] \subseteq X$ and (X, d) is geodesic. It follows that (X, d) is a ℝ-tree by Lemma 3.36.   □

Example 3.41. Let $\mathcal{P}$ denote the collection of partitions of the positive integers ℕ. There is a natural partial order $\leq$ on $\mathcal{P}$ defined by $P \leq Q$ if every block of Q is a subset of some block of P (that is, the blocks of P are unions of blocks of Q). Thus, the partition $\{\{1\}, \{2\}, \ldots\}$ consisting of singletons is the unique largest element of $\mathcal{P}$, while the partition $\{\{1, 2, \ldots\}\}$ consisting of a single block is the unique smallest element. Consider a function $\Pi : \mathbb{R}_+ \mapsto \mathcal{P}$ that is non-increasing in this partial order. Suppose that $\Pi(0) = \{\{1\}, \{2\}, \ldots\}$ and $\Pi(t) = \{\{1, 2, \ldots\}\}$ for all t sufficiently large. Suppose also that if Π is right-continuous in the sense that if i and j don't belong to the same block of $\Pi(t)$ for some $t \in \mathbb{R}_+$, then they don't belong to the same block of $\Pi(u)$ for $u > t$ sufficiently close to t.

Let T denote the set consisting of points of the form (t, B), where $t \in \mathbb{R}_+$ and $B \in \Pi(t)$. Given two point $(s, A), (t, B) \in T$, set

$$m((s, A), (t, B))$$
$$:= \inf\{u > s \wedge t : A \text{ and } B \text{ subsets of a common block of } \Pi(u)\},$$

and put

$$d((s, A), (t, B)) := [m((s, A), (t, B)) - s] + [m((s, A), (t, B)) - t].$$

It is not difficult to check that d is a metric that satisfies the four point condition and that the space T is connected. Hence, (T, d) is a ℝ-tree by Theorem 3.40. The analogue of this construction with ℕ replaced by $\{1, 2, 3, 4\}$ is shown in Figure 3.13.

Moreover, if we let $\bar{T}$ denote the completion of T with respect to the metric d, then $\bar{T}$ is also a ℝ-tree. It is straightforward to check that $\bar{T}$ is compact if and only if $\Pi(t)$ has finitely many blocks for all $t > 0$.

Write δ for the restriction of d to the positive integers ℕ, so that

$$\delta(i, j) = 2 \inf\{t > 0 : i \text{ and } j \text{ belong to the same block of } \Pi(t)\}.$$

The completion $\mathbb{S}$ of ℕ with respect to δ is isometric to the closure of ℕ in $\bar{T}$, and $\mathbb{S}$ is compact if and only if $\Pi(t)$ has finitely many blocks for all $t > 0$. Note that δ is an *ultrametric*, that is, $\delta(x, y) \leq \delta(x, z) \vee \delta(z, y)$ for $x, y, z \in \mathbb{S}$. This implies that at least two of the distances are equal and are no smaller than the third. Hence, all triangles are *isosceles*. When $\mathbb{S}$ is compact, the open balls for the metric δ coincide with the closed balls and are obtained by taking the closure of the blocks of $\Pi(t)$ for $t > 0$. In particular, $\mathbb{S}$ is *totally disconnected*.

The correspondence between coalescing partitions, tree structures and ultrametrics is a familiar idea in the physics literature – see, for example, [109].

3.4 ℝ–Trees without Leaves

3.4.1 Ends

Definition 3.42. *An* ℝ-tree without leaves *is a* ℝ–trees (T, d) *that satisfies the following extra axioms.*

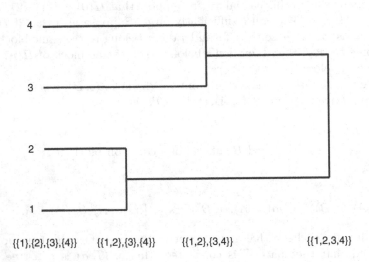

Fig. 3.13. The construction of a ℝ-tree from a non-increasing function taking values in the partitions of $\{1, 2, 3, 4\}$.

Axiom (c) The metric space (T, d) is complete.
Axiom (d) For each $x \in T$ there is at least one isometric embedding $\theta : \mathbb{R} \to T$
 with $x \in \theta(\mathbb{R})$.

Example 3.43. "The" real tree $(\mathcal{T}, D)$ of Example 3.18 satisfies Axioms (c) and (d).

We will suppose in this section that we are always working with a ℝ-tree (T, d) that is without leaves.

Definition 3.44. *An end of T is an equivalence class of isometric embeddings from $\mathbb{R}_+$ into T, where we regard two such embeddings ϕ and ψ as being equivalent if there exist $\alpha \in \mathbb{R}$ and $\beta \in \mathbb{R}_+$ such that $\alpha + \beta \geq 0$ and $\phi(t) = \psi(t + \alpha)$ for all $t \geq \beta$. Write E for the set of ends of T.*

By Axiom (d), E has at least 2 points. Fix a distinguished element † of E. For each $x \in T$ there is a unique isometric embedding $\kappa_x : \mathbb{R}_+ \to T$ such that $\kappa_x(0) = x$ and κ_x is a representative of the equivalence class of †. Similarly, for each $\xi \in E_+ := E \backslash \{†\}$ there is at least one isometric embedding $\theta : \mathbb{R} \to T$ such that $t \mapsto \theta(t)$, $t \geq 0$, is a representative of the equivalence class of ξ

and $t \mapsto \theta(-t)$, $t \geqslant 0$, is a representative of the equivalence class of †. Denote the collection of all such embeddings by Θ_ξ. If $\theta, \theta' \in \Theta_\xi$, then there exists $\gamma \in \mathbb{R}$ such that $\theta(t) = \theta'(t + \gamma)$ for all $t \in \mathbb{R}$. Thus, it is possible to select an embedding $\theta_\xi \in \Theta_\xi$ for each $\xi \in E_+$ in such a way that for any pair $\xi, \zeta \in E_+$ there exists t_0 (depending on ξ, ζ) such that $\theta_\xi(t) = \theta_\zeta(t)$ for all $t \leqslant t_0$ (and $\theta_\xi(]t_0, \infty[) \cap \theta_\zeta(]t_0, \infty[) = \varnothing$). Extend θ_ξ to $\mathbb{R}^* := \mathbb{R} \cup \{\pm\infty\}$ by setting $\theta_\xi(-\infty) := \dagger$ and $\theta_\xi(+\infty) := \xi$.

Example 3.45. The ends of the real tree $(\mathcal{T}, D)$ of Example 3.18 can be identified with the collection consisting of the empty set and the elements of $\mathcal{E}_+$, where $\mathcal{E}_+$ consists of subsets $B \subset \mathbb{R}$ such that $-\infty < \inf B$ and $\sup B = +\infty$. If we choose † to be the empty set so that $\mathcal{E}_+$ plays the role of E_+, then we can define the isometric embedding θ_A for $A \in \mathcal{E}_+$ by $\theta_A(t) := (A \cap] - \infty, t]) \cup \{t\} = A|t$, in the notation of Example 3.18.

The map $(t, \xi) \mapsto \theta_\xi(t)$ from $\mathbb{R} \times E_+$ (resp. $\mathbb{R}^* \times E_+$) into T (resp. $T \cup E$) is surjective. Moreover, if $\eta \in T \cup E$ is in $\theta_\xi(\mathbb{R}^*) \cap \theta_\zeta(\mathbb{R}^*)$ for $\xi, \zeta \in E_+$, then $\theta_\xi^{-1}(\eta) = \theta_\zeta^{-1}(\eta)$. Denote this common value by $h(\eta)$, the *height* of η. In genealogical terminology, we think of $h(\eta)$ as the *generation* to which η belongs. In particular, $h(\dagger) := -\infty$ and $h(\xi) := +\infty$ for $\xi \in E_+$. For the real tree $(\mathcal{T}, D)$ of Example 3.18 with corresponding isometric embeddings defined as above, $h(B)$ is just $\sup B$, with the usual convention that $\sup \varnothing := -\infty$ (in accord with the notation of Example 3.18).

Define a *partial order* $\leqslant$ on $T \cup E$ by declaring that $\eta \leqslant \rho$ if there exists $-\infty \leqslant s \leqslant t \leqslant +\infty$ and $\xi \in E_+$ such that $\eta = \theta_\xi(s)$ and $\rho = \theta_\xi(t)$. In genealogical terminology, $\eta \leqslant \rho$ corresponds to η being an ancestor of ρ (note that individuals are their own ancestors). In particular, † is the unique point that is an ancestor of everybody, while points of E_+ are characterized by being only ancestors of themselves. For the real tree $(\mathcal{T}, D)$ of Example 3.18, $A \leqslant B$ if and only if $A = (B \cap] - \infty, \sup A]) \cup \{\sup A\}$. In particular, this partial order is **not** the usual inclusion partial order (for example, the singleton $\{0\}$ is an ancestor of the singleton $\{1\}$).

Each pair $\eta, \rho \in T \cup E$ has a well-defined *greatest common lower bound* $\eta \wedge \rho$ in this partial order, with $\eta \wedge \rho \in T$ unless $\eta = \rho \in E_+$, $\eta = \dagger$ or $\rho = \dagger$. In genealogical terminology, $\eta \wedge \rho$ is the *most recent common ancestor* of η and ρ. For $x, y \in T$ we have

$$\begin{aligned} d(x, y) &= h(x) + h(y) - 2h(x \wedge y) \\ &= [h(x) - h(x \wedge y)] + [h(y) - h(x \wedge y)]. \end{aligned} \qquad (3.1)$$

Therefore, $h(x) = d(x, y) - h(y) + 2h(x \wedge y) \leqslant d(x, y) + h(y)$ and, similarly, $h(y) \leqslant d(x, y) + h(x)$, so that

$$|h(x) - h(y)| \leqslant d(x, y), \qquad (3.2)$$

with equality if $x, y \in T$ are comparable in the partial order (that is, if $x \leqslant y$ or $y \leqslant x$).

If $x, x' \in T$ are such that $h(x \wedge y) = h(x' \wedge y)$ for all $y \in T$, then, by (3.1), $d(x, x') = [h(x) - h(x \wedge x')] + [h(x') - h(x \wedge x')] = [h(x) - h(x \wedge x)] + [h(x') - h(x' \wedge x')] = 0$, so that $x = x'$. Slight elaborations of this argument show that if $\eta, \eta' \in T \cup E$ are such that $h(\eta \wedge y) = h(\eta' \wedge y)$ for all y in some dense subset of T, then $\eta = \eta'$.

For $x, x', z \in T$ we have that if $h(x \wedge z) < h(x' \wedge z)$, then $x \wedge x' = x \wedge z$ and a similar conclusion holds with the roles of x and x' reversed; whereas if $h(x \wedge z) = h(x' \wedge z)$, then $x \wedge z = x' \wedge z \leqslant x \wedge x'$. Using (3.1) and (3.2) and checking the various cases we find that

$$|h(x \wedge z) - h(x' \wedge z)| \leqslant d(x \wedge z, x' \wedge z) \leqslant d(x, x'). \tag{3.3}$$

For $\eta \in T \cup E$ and $t \in \mathbb{R}^*$ with $t \leqslant h(\eta)$, let $\eta|t$ denote the unique $\rho \in T \cup E$ with $\rho \leqslant \eta$ and $h(\rho) = t$. Equivalently, if $\eta = \theta_\xi(u)$ for some $u \in \mathbb{R}^*$ and $\xi \in E_+$, then $\eta|t = \theta_\xi(t)$ for $t \leqslant u$. For the real tree of Example 3.18, this definition coincides with the one given in Example 3.18.

The metric space (E_+, δ), where

$$\delta(\xi, \zeta) := 2^{-h(\xi \wedge \zeta)},$$

is complete. Moreover, the metric δ is actually an *ultrametric*; that is, $\delta(\xi, \zeta) \leqslant \delta(\xi, \eta) \vee \delta(\eta, \zeta)$ for all $\xi, \zeta, \eta \in E_+$.

3.4.2 The Ends Compactification

Suppose in this subsection that the metric space (E_+, δ) is separable. For $t \in \mathbb{R}$ consider the set

$$T_t := \{x \in T : h(x) = t\} = \{\xi|t : \xi \in E_+\} \tag{3.4}$$

of points in T that have height t. For each $x \in T_t$ the set $\{\zeta \in E_+ : \zeta|t = x\}$ is a ball in E_+ of diameter at most 2^{-t} and two such balls are disjoint. Thus, the separability of E_+ is equivalent to each of the sets T_t being countable. In particular, separability of E_+ implies that T is also separable, with countable dense set $\{\xi|t : \xi \in E_+, t \in \mathbb{Q}\}$, say.

We can, via a standard Stone–Čech-like procedure, embed $T \cup E$ in a compact metric space in such a way that for each $y \in T \cup E$ the map $x \mapsto h(x \wedge y)$ has a continuous extension to the compactification (as an extended real–valued function).

More specifically, let S be a countable dense subset of T. Let π be a strictly increasing, continuous function that maps $\mathbb{R}$ onto $]0, 1[$. Define an injective map Π from T into the compact, metrizable space $[0, 1]^S$ by $\Pi(x) := (\pi(h(x \wedge y)))_{y \in S}$. Identify T with $\Pi(T)$ and write $\overline{T}$ for the closure of $T(= \Pi(T))$ in $[0, 1]^T$. In other words, a sequence $\{x_n\}_{n \in \mathbb{N}} \subset T$ converges to a point in $\overline{T}$ if $h(x_n \wedge y)$ converges (possibly to $-\infty$) for all $y \in S$, and two such sequences $\{x_n\}_{n \in \mathbb{N}}$ and $\{x'_n\}_{n \in \mathbb{N}}$ converge to the same point if and only if $\lim_n h(x_n \wedge y) = \lim_n h(x'_n \wedge y)$ for all $y \in S$.

We can identify distinct points in $T \cup E$ with distinct points in $\overline{T}$. If $\{x_n\}_{n \in \mathbb{N}} \subset T$ and $\xi \in E_+$ are such that for all $t \in \mathbb{R}$ we have $\xi | t \leqslant x_n$ for all sufficiently large n, then $\lim_n h(x_n \wedge y) = h(\xi \wedge y)$ for all $y \in S$. We leave the identification of † to the reader.

In fact, we have $\overline{T} = T \cup E$. To see this, suppose that $\{x_n\}_{n \in \mathbb{N}} \subset T$ converges to $x_\infty \in \overline{T}$. Put $h_\infty := \sup_{y \in S} \lim_n h(x_n \wedge y)$. Assume for the moment that $h_\infty \in \mathbb{R}$. We will show that $x_\infty \in T$ with $h(x_\infty) = h_\infty$. For all $k \in \mathbb{N}$ we can find $y_k \in S$ such that

$$h_\infty - \frac{1}{k} \leqslant \lim_n h(x_n \wedge y_k) \leqslant h(y_k) \leqslant h_\infty + \frac{1}{k}.$$

Observe that

$$d(y_k, y_\ell) \leqslant \limsup_n \Big(d(y_k, x_n \wedge y_k) + d(x_n \wedge y_k, x_n \wedge y_\ell)$$

$$+ d(x_n \wedge y_\ell, y_\ell) \Big)$$

$$= \limsup_n \Big([h(y_k) - h(x_n \wedge y_k)] + |h(x_n \wedge y_k) - h(x_n \wedge y_\ell)|$$

$$+ [h(y_\ell) - h(x_n \wedge y_\ell)] \Big)$$

$$\leqslant \frac{2}{k} + \left(\frac{1}{k} + \frac{1}{\ell} \right) + \frac{2}{\ell}.$$

Therefore, $(y_k)_{k \in \mathbb{N}}$ is a d-Cauchy sequence and, by Axiom (c), this sequence converges to $y_\infty \in T$. Moreover, by (3.2) and (3.3), $\lim_n h(x_n \wedge y_\infty) = h(y_\infty) = h_\infty$.

We claim that $y_\infty = x_\infty$; that is, $\lim_n h(x_n \wedge z) = h(y_\infty \wedge z)$ for all $z \in S$. To see this, fix $z \in T$ and $\epsilon > 0$. If n is sufficiently large, then

$$h(x_n \wedge z) \leqslant h(y_\infty) + \epsilon \tag{3.5}$$

and

$$h(y_\infty) - \epsilon \leqslant h(x_n \wedge y_\infty) \leqslant h(y_\infty). \tag{3.6}$$

If $h(y_\infty \wedge z) \leqslant h(y_\infty) - \epsilon$, then (3.6) implies that $y_\infty \wedge z = x_n \wedge z$. On the other hand, if $h(y_\infty \wedge z) \geqslant h(y_\infty) - \epsilon$, then (3.6) implies that

$$h(x_n \wedge z) \geqslant h(y_\infty) - \epsilon, \tag{3.7}$$

and so, by (3.5) and (3.6),

$$|h(y_\infty \wedge z) - h(x_n, z)|$$

$$\leqslant [h(y_\infty) - (h(y_\infty) - \epsilon)] \vee [(h(y_\infty) + \epsilon) - (h(y_\infty) - \epsilon)] \tag{3.8}$$

$$= 2\epsilon.$$

We leave the analogous arguments for $h_\infty = +\infty$ (in which case $x_\infty \in E_+$) and $h_\infty = -\infty$ (in which case $x_\infty = †$) to the reader.

We have just seen that the construction of $\overline{T}$ does not depend on T (more precisely, any two such compactifications are homeomorphic). Moreover, a sequence $\{x_n\}_{n\in\mathbb{N}} \subset T \cup E$ converges to a limit in $T \cup E$ if and only if $\lim_n h(x_n \wedge y)$ exists for all $y \in T$, and two convergent sequences $\{x_n\}_{n\in\mathbb{N}}$ and $\{x'_n\}_{n\in\mathbb{N}}$ converge to the same limit if and only if $\lim_n h(x_n \wedge y) = \lim_n h(x'_n \wedge y)$ for all $y \in T$.

3.4.3 Examples of ℝ-Trees without Leaves

Fix a prime number p and constants $r_-, r_+ \geqslant 1$. Let $\mathbb{Q}$ denote the rational numbers. Define an equivalence relation $\sim$ on $\mathbb{Q} \times \mathbb{R}$ as follows. Given $a, b \in \mathbb{Q}$ with $a \neq b$ write $a - b = p^{v(a,b)}(m/n)$ for some $v(a,b), m, n \in \mathbb{Z}$ with m and n not divisible by p. For $v(a,b) \geqslant 0$ put $w(a,b) = \sum_{i=0}^{v(a,b)} r_+^i$, and for $v(a,b) < 0$ put $w(a,b) := 1 - \sum_{i=0}^{-v(a,b)} r_-^i$. Set $w(a,a) := +\infty$. Given $(a,s), (b,t) \in \mathbb{Q} \times \mathbb{R}$ declare that $(a,s) \sim (b,t)$ if and only if $s = t \leqslant w(a,b)$. Note that

$$v(a,c) \geqslant v(a,b) \wedge v(b,c) \tag{3.9}$$

so that

$$w(a,c) \geqslant w(a,b) \wedge w(b,c) \tag{3.10}$$

and $\sim$ is certainly transitive (reflexivity and symmetry are obvious).

Let T denote the collection of equivalence classes for this equivalence relation. Define a partial order $\leqslant$ on T as follows. Suppose that $x, y \in T$ are equivalence classes with representatives (a,s) and (b,t). Say that $x \leqslant y$ if and only if $s \leqslant w(a,b) \wedge t$. It follows from (3.10) that $\leqslant$ is indeed a partial order. A pair $x, y \in T$ with representatives (a,s) and (b,t) has a unique greatest common lower bound $x \wedge y$ in this order given by the equivalence class of $(a, s \wedge t \wedge w(a,b))$, which is also the equivalence class of $(b, s \wedge t \wedge w(a,b))$.

For $x \in T$ with representative (a,s), put $h(x) := s$. Define a metric d on T by setting $d(x,y) := h(x) + h(y) - 2h(x \wedge y)$. We leave it to the reader to check that (T,d) is a ℝ–tree satisfying Axioms (a)–(d), and that the definitions of $x \leqslant y$, $x \wedge y$ and $h(x)$ fit into the general framework of Section 3.4, with the set E_+ corresponding to $\mathbb{Q} \times \mathbb{R}$–valued paths $s \mapsto (a(s),s)$ such that $s \leqslant w(a(s),a(t)) \wedge t$.

Note that there is a natural Abelian group structure on E_+: if ξ and ζ correspond to paths $s \mapsto (a(s),s)$ and $s \mapsto (b(s),s)$, then define $\xi + \zeta$ to correspond to the path $s \mapsto (a(s) + b(s),s)$. We mention in passing that there is a bi–continuous group isomorphism between E_+ and the additive group of the p–adic integers $\mathbb{Q}_p$. (This map is, however, not an isometry if E_+ is equipped with the δ metric and $\mathbb{Q}_p$ is equipped with the usual p-adic metric.)

4

Hausdorff and Gromov–Hausdorff Distance

4.1 Hausdorff Distance

We follow the presentation in [37] in this section and omit some of the more elementary proofs.

Definition 4.1. *Denote by $U_r(S)$ the r-neighborhood of a set S in a metric space (X, d). That is, $U_r(S) := \{x \in X : d(x, S) < r\}$, where $d(x, S) := \inf\{d(x, y) : y \in S\}$. Equivalently, $U_r(S) := \bigcup_{x \in S} B_r(x)$, where $B_r(x)$ is the open ball of radius r centered at x.*

Definition 4.2. *Let A and B be subsets of a metric space (X, d). The Hausdorff distance between A and B, denoted by $d_H(A, B)$, is defined by*

$$d_H(A, B) := \inf\{r > 0 : A \subset U_r(B) \text{ and } B \subset U_r(A)\}.$$

See Figure 4.1

Proposition 4.3. *Let (X, d) be a metric space. Then*

(i) d_H is a semi-metric on the set of all subsets of X.
(ii) $d_H(A, \overline{A}) = 0$ for any $A \subseteq X$, where $\overline{A}$ denotes the closure of A.
(iii) If A and B are closed subsets of X and $d_H(A, B) = 0$, then $A = B$.

Let $\mathfrak{M}(X)$ denote the set of non-empty closed subsets of X equipped with Hausdorff distance. Proposition 4.3 says that $\mathfrak{M}(X)$ is a metric space (provided we allow the metric to take the value $+\infty$).

Proposition 4.4. *If the metric space (X, d) is complete, then the metric space $(\mathfrak{M}(X), d_H)$ is also complete.*

Proof. Consider a Cauchy sequence $\{S_n\}_{n \in \mathbb{N}}$ in $\mathfrak{M}(X)$. Let S denote the set of points $x \in X$ such that any neighborhood of x intersects with infinitely many of the S_n. That is, $S := \bigcap_{m=1}^{\infty} \overline{\bigcup_{n=m}^{\infty} S_n}$. By definition of the Hausdorff metric,

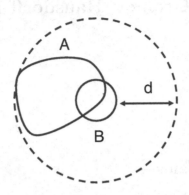

Fig. 4.1. The Hausdorff distance between the sets A and B is d

we can find a sequence $\{y_n\}_{n \in \mathbb{N}}$ such that $y_n \in S_n$ and $d(y_m, y_n) \leqslant d_H(S_m, S_n)$ for all $m, n \in \mathbb{N}$. Since X is complete, $\lim_{n \to} y_n = y$ exists. Note that $y \in S$ and so S is non-empty. By definition, S is closed, and so $S \in \mathfrak{M}(X)$.

We will show that

$$S_n \to S.$$

Fix $\epsilon > 0$ and let n_0 be such that $d_H(S_n, S_m) < \epsilon$ for all $m, n \geqslant n_0$. It suffices to show that $d_H(S, S_n) < 2\epsilon$ for any $n \geqslant n_0$, and this is equivalent to showing that:

$$\text{For } x \in S \text{ and } n \geqslant n_0, \ d(x, S_n) < 2\epsilon. \tag{4.1}$$

$$\text{For } x \in S_n \text{ and } n \geqslant n_0, \ d(x, S) < 2\epsilon. \tag{4.2}$$

To establish (4.1), note first that there exists an $m \geqslant n_0$ such that $B_\epsilon(x) \cap S_m \neq \varnothing$. In other words, there is a point $y \in S_m$ such that $d(x, y) < \epsilon$. Since $d_H(S_n, S_m) < \epsilon$, we also have $d(y, S_n) < \epsilon$, and, therefore, $d(x, S_n) < 2\epsilon$.

Turning to (4.2), let $n_1 = n$ and for every integer $k > 1$ choose an index n_k such that $n_k > n_{k+1}$ and $d_H(S_p, S_q) < \epsilon/2^k$ for all $p, q \geqslant n_k$. Define a sequence of points $\{x_k\}_{k \in \mathbb{N}}$, where $x_k \in S_{n_k}$, as follows: let $x_1 = x$, and x_{k+1} be a point of $S_{n_{k+1}}$ such that $d(x_k, x_{k+1}) < \epsilon/2^k$ for all k. Such a point can be found because $d_H(S_{n_k}, S_{n_{k+1}}) < \epsilon/2^k$.

Since $\sum_{k\in\mathbb{N}} d(x_k, x_{k+1}) < 2\epsilon < \infty$, the sequence $\{x_k\}_{k\in\mathbb{N}}$ is a Cauchy sequence. Hence, it converges to a point $y \in X$ by the assumed completeness of X. Then,

$$d(x, y) = \lim_{n\to\infty} d(x, x_n) \leqslant \sum_{k\in\mathbb{N}} d(x_k, x_{k+1}) < 2\epsilon.$$

Because $y \in S$ by construction, it follows that $d(x, S) < 2\epsilon$.   □

Theorem 4.5. *If the metric space (X, d) is compact, then the metric space $(\mathfrak{M}(X), d_H)$ is also compact.*

Proof. By Proposition 4.4, $\mathfrak{M}(X)$ is complete. Therefore, it suffices to prove that $\mathfrak{M}(X)$ is totally bounded. Let S be a finite ϵ-net in X. We will show that the set of all non-empty subsets of S is an ϵ-net in $\mathfrak{M}(X)$.

Let $A \in \mathfrak{M}(X)$. Consider

$$S_A = \{x \in S : d(x, A) \leqslant \epsilon\}.$$

Since S is an ϵ-net in X, for every $y \in A$ there exists an $x \in S$ such that $d(x, y) \leqslant \epsilon$. Because $d(x, A) \leqslant d(x, y) \leqslant \epsilon$, this point x belongs to S_A. Therefore, $d(y, S_A) \leqslant \epsilon$ for all $y \in A$.

Since $d(x, A) \leqslant \epsilon$ for any $x \in S_A$, it follows that $d_H(A, S_A) \leqslant \epsilon$. Since A is arbitrary, this proves that the set of subsets of S is an ϵ-net in $\mathfrak{M}(X)$.   □

4.2 Gromov–Hausdorff Distance

In this section we follow the development in [37]. Similar treatments may be found in [80, 34].

4.2.1 Definition and Elementary Properties

Definition 4.6. *Let X and Y be metric spaces. The* Gromov–Hausdorff *distance between them, denoted by $d_{\mathrm{GH}}(X, Y)$, is the infimum of the Hausdorff distances $d_H(X', Y')$ over all metric spaces Z and subspaces X' and Y' of Z that are isometric to X and Y, respectively – see Figure 4.2.*

Remark 4.7. It is not necessary to consider all possible embedding spaces Z. The Gromov–Hausdorff distance between two metric spaces (X, d_X) and (Y, d_Y) is the infimum of those $r > 0$ such that there exists a metric d on the disjoint union $X \bigsqcup Y$ such that the restrictions of d to X and Y coincide with d_X and d_Y and $d_H(X, Y) < r$ in the space $(X \bigsqcup Y, d)$.

Proposition 4.8. *The distance d_{GH} satisfies the triangle inequality.*

Proof. Given d_{XY} on $X \bigsqcup Y$ and d_{YZ} on $Y \bigsqcup Z$, define d_{XZ} on $X \bigsqcup Z$ by

$$d_{XZ}(x, z) = \inf_{y\in Y} \{d_{XY}(x, y) + d_{YZ}(y, z)\}.$$

□

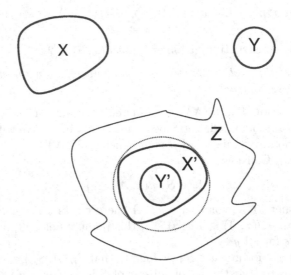

Fig. 4.2. Computation of the Gromov–Hausdorff distance between metric spaces X and Y by embedding isometric copies X' and Y' into Z

4.2.2 Correspondences and ϵ-Isometries

The definition of the Gromov–Hausdorff distance $d_{\mathrm{GH}}(X, Y)$ is somewhat unwieldy, as it involves an infimum over metric spaces Z and isometric embeddings of X and Y in Z. Remark 4.7 shows that it is enough to take Z to be the disjoint union of X and Y, but this still leaves the problem of finding optimal metrics on the disjoint union that extend the metrics on X and Y. In this subsection we will give a more effective formulation of the Gromov–Hausdorff distance, as well as convenient upper and lower bounds on the distance.

Definition 4.9. *Let X and Y be two sets. A* correspondence *between X and Y is a set $\mathfrak{R} \subset X \times Y$ such that for every $x \in X$ there exists at least one $y \in Y$ for which $(x, y) \in \mathfrak{R}$, and similarly for every $y \in Y$ there exists an $x \in X$ for which $(x, y) \in \mathfrak{R}$ – see Figure 4.3.*

Definition 4.10. *Let $\mathfrak{R}$ be a correspondence between metric spaces X and Y. The* distortion *of $\mathfrak{R}$ is defined to be*

$$\mathrm{dis}\,\mathfrak{R} := \sup\{|d_X(x, x') - d_Y(y, y')| : (x, y), (x', y') \in \mathfrak{R}\},$$

where d_X and d_Y are the metrics of X and Y respectively.

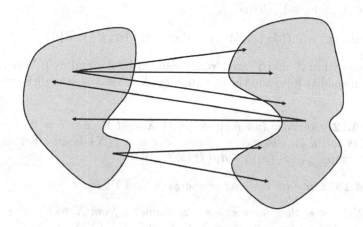

Fig. 4.3. A correspondence between two spaces

Theorem 4.11. *For any two metric spaces X and Y,*

$$d_{\text{GH}}(X, Y) = \frac{1}{2} \inf_{\mathfrak{R}} (\text{dis}\mathfrak{R})$$

where the infimum is taken over all correspondences $\mathfrak{R}$ between X and Y.

Proof. We first show for any $r > d_{\text{GH}}(X, Y)$ that there exists a correspondence $\mathfrak{R}$ with $\text{dis}\mathfrak{R} < 2r$. Indeed, since $d_{\text{GH}}(X, Y) < r$, we may assume that X and Y are subspaces of some metric space Z and $d_H(X, Y) < r$ in Z. Define

$$\mathfrak{R} = \{(x, y) : x \in X, y \in Y, d(x, y) < r\}$$

where d is the metric of Z.

That $\mathfrak{R}$ is a correspondence follows from the fact that $d_H(X, Y) < r$. The estimate $\text{dis}\mathfrak{R} < 2r$ follows from the triangle inequality: if $(x, y) \in \mathfrak{R}$ and $(x', y') \in \mathfrak{R}$, then

$$|d(x, x') - d(y, y')| \leqslant d(x, y) + d(x', y') < 2r.$$

Conversely, we show that $d_{\text{GH}}(X, Y) \leqslant \frac{1}{2}\text{dis}\mathfrak{R}$ for any correspondence $\mathfrak{R}$. Let $\text{dis}\mathfrak{R} = 2r$. To avoid confusion, we use the notation d_X and d_Y for the metrics of X, Y, respectively. It suffices to show that there is a metric d on the disjoint union $X \bigsqcup Y$ such that $d|_{X \times X} = d_X$, $d|_{Y \times Y} = d_Y$, and $d_H(X, Y) \leqslant r$ in $(X \bigsqcup Y, d)$.

Given $x \in X$ and $y \in Y$, define

$$d(x, y) = \inf\{d_X(x, x') + r + d_Y(y', y) : (x', y') \in \mathfrak{R}\}$$

(the distances within X and Y are already defined by d_X and d_Y). Verifying the triangle inequality for d and the fact that $d_H(X, Y) \leqslant r$ is straightforward.

□

Definition 4.12. *Consider two metric spaces X and Y. For $\epsilon > 0$, a map $f : X \to Y$ is called an ϵ-isometry if $\mathrm{dis} f \leqslant \epsilon$ and $f(X)$ is an ϵ-net in Y. (Here $\mathrm{dis} f := \sup_{x,y \in X} |d_X(x, y) - d_Y(f(x), f(y))|$.)*

Corollary 4.13. *Consider two metric spaces X and Y. Fix $\epsilon > 0$.*

(i) If $d_{\mathrm{GH}}(X, Y) < \epsilon$, then there exists a 2ϵ-isometry from X to Y.
(ii) If there exists an ϵ-isometry from X to Y, then $d_{\mathrm{GH}}(X, Y) < 2\epsilon$.

Proof. (i) Let $\mathfrak{R}$ be a correspondence between X and Y with $\mathrm{dis}\mathfrak{R} < 2\epsilon$. For every $x \in X$, choose $f(x) \in Y$ such that $(x, f(x)) \in \mathfrak{R}$. This defines a map $f : X \to Y$. Obviously $\mathrm{dis} f \leqslant \mathrm{dis}\mathfrak{R} < 2\epsilon$. We will show that $f(X)$ is an ϵ-net in Y.

For a $y \in Y$, consider an $x \in X$ such that $(x, y) \in \mathfrak{R}$. Since both y and $f(x)$ are in correspondence with x, it follows that $d(y, f(x)) \leqslant d(x, x) + \mathrm{dis}\mathfrak{R} < 2\epsilon$. Hence, $d(y, f(X)) < 2\epsilon$.

(ii) Let f be an ϵ-isometry. Define $\mathfrak{R} \subset X \times Y$ by

$$\mathfrak{R} = \{(x, y) \in X \times Y : d(y, f(x)) \leqslant \epsilon\}.$$

Then $\mathfrak{R}$ is a correspondence because $f(X)$ is an ϵ-net in Y. If $(x, y) \in \mathfrak{R}$ and $(x', y') \in \mathfrak{R}$, then

$$|d(y, y') - d(x, x')| \leqslant |d(f(x), f(x')) - d(x, x')| + d(y, f(x)) + d(y', f(x'))$$
$$\leqslant \mathrm{dis} f + \epsilon + \epsilon \leqslant 3\epsilon.$$

Hence, $\mathrm{dis}\mathfrak{R} < 3\epsilon$, and Theorem 4.11 implies

$$d_{\mathrm{GH}}(X, Y) \leqslant \frac{3}{2}\epsilon < 2\epsilon.$$

□

4.2.3 Gromov–Hausdorff Distance for Compact Spaces

Theorem 4.14. *The Gromov–Hausdorff distance is a metric on the space of isometry classes of compact metric spaces.*

Proof. We already know that d_{GH} is a semi-metric, so only that show $d_{\mathrm{GH}}(X,Y) = 0$ implies that X and Y are isometric.

Let X and Y be two compact spaces such that $d_{\mathrm{GH}}(X,Y) = 0$. By Corollary 4.13, there exists a sequence of maps $f_n : X \to Y$ such that $\mathrm{dis}\, f_n \to 0$.

Fix a countable dense set $S \subset X$. Using Cantor's diagonal procedure, choose a subsequence $\{f_{n_k}\}$ of $\{f_n\}$ such that for every $x \in S$ the sequence $\{f_{n_k}(x)\}$ converges in Y. By renumbering, we may assume that this holds for $\{f_n\}$ itself. Define a map $f : S \to Y$ as the limit of the f_n, namely, set $f(x) = \lim f_n(x)$ for every $x \in S$.

Because

$$|d(f_n(x), f_n(y)) - d(x,y)| \leqslant \mathrm{dis}\, f_n \to 0,$$

we have

$$d(f(x), f(y)) = \lim d(f_n(x), f_n(y)) = d(x,y) \quad \text{for all } x, y \in S.$$

In other words, f is a distance-preserving map from S to Y. Then f can be extended to a distance-preserving map from X to Y. Now interchange the roles of X and Y. $\qquad\square$

Proposition 4.15. *Consider compact metric spaces X and $\{X_n\}_{n \in \mathbb{N}}$. The sequence $\{X_n\}_{n \in \mathbb{N}}$ converges to X in the Gromov–Hausdorff distance if and only if for every $\epsilon > 0$ there exists a finite ϵ-net S in X and an ϵ-net S_n in each X_n such that S_n converges to S in the Gromov–Hausdorff distance.*

Moreover these ϵ-nets can be chosen so that, for all sufficiently large n, S_n has the same cardinality as S.

Definition 4.16. *A collection $\mathfrak{X}$ of compact metric spaces is* uniformly totally bounded *if for every $\epsilon > 0$ there exists a natural number $N = N(\epsilon)$ such that every $X \in \mathfrak{X}$ contains an ϵ-net consisting of no more than N points.*

Remark 4.17. Note that if the collection $\mathfrak{X}$ of compact metric spaces is uniformly totally bounded, then there is a constant D such that $\mathrm{diam}(X) \leqslant D$ for all $X \in \mathfrak{X}$.

Theorem 4.18. *A uniformly totally bounded class $\mathfrak{X}$ of compact metric spaces is pre-compact in the Gromov–Hausdorff topology.*

Proof. Let $N(\epsilon)$ be as in Definition 4.16 and D be as in Remark 4.17. Define $N_1 = N(1)$ and $N_k = N_{k-1} + N(1/k)$ for all $k \geqslant 2$. Let $\{X_n\}_{n \in \mathbb{N}}$ be a sequence of metric spaces from $\mathfrak{X}$.

In every space X_n, consider a union of $(1/k)$-nets for all $k \in \mathbb{N}$. This is a countable dense collection $S_n = \{x_{i,n}\}_{i \in \mathbb{N}} \subset X_n$ such that for every k the first N_k points of S_n form a $(1/k)$-net in X_n. The distances $d_{X_n}(x_{i,n}, x_{j,n})$ do not exceed D, i.e. belong to a compact interval. Therefore, using the Cantor diagonal procedure, we can extract a subsequence of $\{X_n\}$ in which $\{d_{X_n}(x_{i,n}, x_{j,n})\}_{n \in \mathbb{N}}$ converge for all i, j. To simplify the notation, we assume that these sequences converge without passing to a subsequence.

We will construct the limit space $\bar{X}$ for $\{X_n\}_{n\in\mathbb{N}}$ as follows. First, pick an abstract countable set $X = \{x_i\}_{i\in\mathbb{N}}$ and define a semi-metric d on X by

$$d(x_i, x_j) = \lim_{n\to\infty} d_{X_n}(x_{n,i}, x_{n,j}).$$

A quotient construction gives us a metric space X/d. We will denote by $\bar{x}_i$ the point of X/d obtained from x_i. Let $\bar{X}$ be the completion of X/d.

For $k \in \mathbb{N}$, consider the set $S^{(k)} = \{\bar{x}_i : 1 \leqslant i \leqslant N_k\} \subset \bar{X}$. Note that $S^{(k)}$ is a $(1/k)$-net in $\bar{X}$. Indeed, every set $S_n^{(k)} = \{\bar{x}_{i,n} : 1 \leqslant i \leqslant N_k\}$ is a $(1/k)$-net in the respective space X_n. Hence, for every $x_{i,n} \in S_n$ there is a $j \leqslant N_k$ such that $d_{X_n}(x_{i,n}, x_{j,n}) \leqslant 1/k$ for infinitely many indices n. Passing to the limit, we see that $d(\bar{x}_i, \bar{x}_j) \leqslant 1/k$ for this j. Thus, $S^{(n)}$ is a $(1/k)$-net in X/d. Hence, $S^{(n)}$ is also a $(1/k)$-net in in $\bar{X}$. Since $\bar{X}$ is complete and has a $(1/k)$-net for any $k \in \mathbb{N}$, $\bar{X}$ is compact.

Furthermore, the set $S^{(k)}$ is a Gromov–Hausdorff limit of the sets $S_n^{(k)}$ as $n \to \infty$, because these are finite sets consisting of N_k points (some of which may coincide) and there is a way of matching up the points of $S_n^{(k)}$ with those in $S^{(k)}$ so that distances converge. Thus, for every $k \in \mathbb{N}$ we have a $(1/k)$-net in $\bar{X}$ that is a Gromov–Hausdorff limit of some $(1/k)$-nets in the spaces X_n. By Proposition 4.15, it follows that X_n converges to $\bar{X}$ in the Gromov–Hausdorff distance. □

4.2.4 Gromov–Hausdorff Distance for Geodesic Spaces

Theorem 4.19. *Let $\{X_n\}_{n\in\mathbb{N}}$ be a sequence of geodesic spaces and X a complete metric space such that X_n converges to X in the Gromov–Hausdorff distance. Then X is a geodesic space.*

Proof. Because X is complete, it suffices to prove that for any two points $x, y \in X$ there is a point $z \in X$ such that $d(x, z) = \frac{1}{2}d(x, y)$ and $d(y, z) = \frac{1}{2}d(x, y)$. Again by completeness, it further suffices to show that for any $\epsilon > 0$ there is a point $z \in X$ such that $|d(x, z) - \frac{1}{2}d(x, y)| < \epsilon$ and $|d(y, z) - \frac{1}{2}d(x, y)| < \epsilon$.

Let n be such that $d_{\mathrm{GH}}(X, X_n) < \epsilon/4$. Then, by Theorem 4.11, there is a correspondence $\mathfrak{R}$ between X and X_n whose distortion is less than $\epsilon/2$. Take points $\tilde{x}, \tilde{y} \in X_n$ corresponding to x and y. Since X_n is a geodesic space, there is a $\tilde{z} \in X_n$ such that $d(\tilde{x}, \tilde{z}) = d(\tilde{z}, \tilde{y}) = \frac{1}{2}d(\tilde{x}, \tilde{y})$. Let $z \in X$ be a point corresponding to $\tilde{z}$. Then

$$\left| d(x, z) - \frac{1}{2}d(x, y) \right| \leqslant \left| d(\tilde{x}, \tilde{z}) - \frac{1}{2}d(\tilde{x}, \tilde{y}) \right| + 2\mathrm{dis}\mathfrak{R} < \epsilon.$$

Similarly, $|d(y, z) - \frac{1}{2}d(x, y)| < \epsilon$. □

Proposition 4.20. *Every compact geodesic space can be obtained as a Gromov–Hausdorff limit of a sequence of finite graphs with edge lengths.*

4.3 Compact ℝ-Trees and the Gromov–Hausdorff Metric

4.3.1 Unrooted ℝ-Trees

Definition 4.21. *Let* $(\mathbf{T}, d_{\mathrm{GH}})$ *be the metric space of isometry classes of compact real trees equipped with the Gromov-Hausdorff metric.*

Lemma 4.22. *The set* $\mathbf{T}$ *of compact ℝ-trees is a closed subset of the space of compact metric spaces equipped with the Gromov-Hausdorff distance.*

Proof. It suffices to note that the limit of a sequence in $\mathbf{T}$ is a geodesic space and satisfies the four point condition.   □

Theorem 4.23. *The metric space* $(\mathbf{T}, d_{\mathrm{GH}})$ *is complete and separable.*

Proof. We start by showing separability. Given a compact ℝ-tree, T, and $\varepsilon > 0$, let S_ε be a finite ε-net in T. Write T_ε for the *subtree of T spanned by* S_ε, that is,

$$T_\varepsilon := \bigcup\nolimits_{x,y \in S_\varepsilon} [x, y] \quad \text{and} \quad d_{T_\varepsilon} := d|_{T_\varepsilon}. \tag{4.3}$$

Obviously, T_ε is still an ε-net for T. Hence, $d_{\mathrm{GH}}(T_\varepsilon, T) \leqslant d_H(T_\varepsilon, T) \leqslant \varepsilon$.

Now each T_ε is just a "finite tree with edge lengths" and can clearly be approximated arbitrarily closely in the d_{GH}-metric by trees with the same tree topology (that is, "shape"), and rational edge lengths. The set of isometry types of finite trees with rational edge lengths is countable, and so $(\mathbf{T}, d_{\mathrm{GH}})$ is separable.

It remains to establish completeness. It suffices by Lemma 4.22 to show that any Cauchy sequence in $\mathbf{T}$ converges to some compact metric space, or, equivalently, any Cauchy sequence in $\mathbf{T}$ has a subsequence that converges to some metric space.

Let $(T_n)_{n \in \mathbb{N}}$ be a Cauchy sequence in $\mathbf{T}$. By Theorem 4.18, a sufficient condition for this sequence to have a subsequential limit is that for every $\varepsilon > 0$ there exists a positive number $N = N(\varepsilon)$ such that every T_n contains an ε-net of cardinality N.

Fix $\varepsilon > 0$ and $n_0 = n_0(\varepsilon)$ such that $d_{\mathrm{GH}}(T_m, T_n) < \varepsilon/2$ for $m, n \geqslant n_0$. Let S_{n_0} be a finite $(\varepsilon/2)$-net for T_{n_0} of cardinality N. Then by (4.11) for each $n \geqslant n_0$ there exists a correspondence $\Re_n$ between T_{n_0} and T_n such that $\mathrm{dis}(\Re_n) < \varepsilon$.

For each $x \in T_{n_0}$, choose $f_n(x) \in T_n$ such that $(x, f_n(x)) \in \Re_n$. Since for any $y \in T_n$ with $(x, y) \in \Re_n$, $d_{T_n}(y, f_n(x)) \leqslant \mathrm{dis}(\Re_n)$, for all $n \geqslant n_0$, the set $f_n(S_{n_0})$ is an ε-net of cardinality N for T_n, $n \geqslant n_0$.   □

4.3.2 Trees with Four Leaves

The following result is elementary, but we include the proof because it includes some formulae that will be useful later.

Lemma 4.24. *The isometry class of a compact $\mathbb{R}$-tree tree (T, d) with four leaves is uniquely determined by the distances between the leaves of T.*

Proof. Let $\{a, b, c, d\}$ be the set of leaves of T. The tree T has one of four possible shapes shown in Figure 4.4.

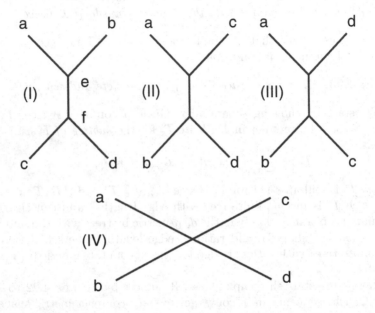

Fig. 4.4. The four leaf-labeled trees with four leaves

Consider case (I), and let e be the uniquely determined branch point on the tree that lies on the segments $[a, b]$ and $[a, c]$, and f be the uniquely determined branch point on the tree that lies on the segments $[c, d]$ and $[a, c]$. That is,

$$e := Y(a, b, c) = Y(a, b, d)$$

and

$$f := Y(c, d, a) = Y(c, d, b).$$

Observe that

$$d(a,e) = \frac{1}{2}(d(a,b) + d(a,c) - d(b,c)) = (b \cdot c)_a = (b \cdot d)_a$$

$$d(b,e) = \frac{1}{2}(d(a,b) + d(b,c) - d(a,c)) = (a \cdot c)_b = (a \cdot d)_b$$

$$d(c,f) = \frac{1}{2}(d(c,d) + d(a,c) - d(a,d)) = (d \cdot a)_c = (d \cdot b)_c \qquad (4.4)$$

$$d(d,f) = \frac{1}{2}(d(c,d) + d(a,d) - d(a,c)) = (c \cdot a)_d = (c \cdot b)_d$$

$$d(e,f) = \frac{1}{2}(d(a,d) + d(b,c) - d(a,b) - d(c,d)) = (a \cdot b)_f = (c \cdot d)_e.$$

Similar observations for the other cases show that if we know the shape of the tree, then we can determine its edge lengths from leaf-to-leaf distances. Note also that

$$\frac{1}{2}\left(d(a,c) + d(b,d) - d(a,b) - d(c,d)\right)$$

$$= \begin{cases} > 0 & \text{for shape (I),} \\ < 0 & \text{for shape (II),} \\ = 0 & \text{for shapes (III) and (IV)} \end{cases} \qquad (4.5)$$

This and analogous inequalities for the quantities that reconstruct the length of the "internal" edge in shapes (II) and (III), respectively, show that the shape of the tree can also be reconstructed from leaf-to-leaf distances. □

4.3.3 Rooted ℝ-Trees

Definition 4.25. *A* rooted ℝ-tree , (X, d, ρ), *is a* ℝ-tree (X, d) *with a distinguished point $\rho \in X$ that we call the* root *. It is helpful to use genealogical terminology and think of ρ as a common ancestor and $h(x) := d(\rho, x)$ as the real-valued generation to which $x \in X$ belongs ($h(x)$ is also called the* height *of x).*

We define a partial order $\leqslant$ on X by declaring that
- *$x \leqslant y$ if $x \in [\rho, y]$, so that x is an ancestor of y.*

Each pair $x, y \in X$ has a well-defined greatest common lower bound*, $x \wedge y$, in this partial order that we think of as the most recent common ancestor of x and y – see Figure 4.5.*

Definition 4.26. *Let $\mathbf{T}^{\mathrm{root}}$ denote the collection of all root-invariant isometry classes of rooted compact ℝ-trees, where we define a root-invariant isometry to be an isometry*

$$\xi : (X_1, d_{X_1}, \rho_1) \to (X_2, d_{X_2}, \rho_2) \text{ with } \xi(\rho_1) = \rho_2.$$

Define the rooted Gromov-Hausdorff distance*, $d_{\mathrm{GH^{root}}}((X_1, \rho_1), (X_2, \rho_2))$, between two rooted ℝ-trees (X_1, ρ_1) and (X_2, ρ_2)*

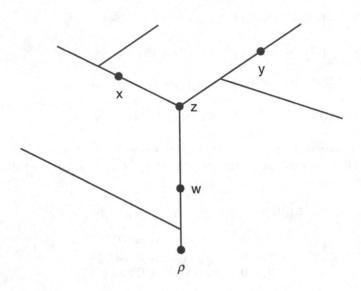

Fig. 4.5. A tree rooted at ρ. Here $w \leqslant x$ and $w \leqslant y$ and also $z \leqslant x$ and $z \leqslant y$. The greatest common lower bound of x and y is z.

as the infimum of $d_{\mathrm{H}}(X'_1, X'_2) \vee d_Z(\rho'_1, \rho'_2)$ over all rooted $\mathbb{R}$-trees (X'_1, ρ'_1) and (X'_2, ρ'_2) that are root-invariant isomorphic to (X_1, ρ_1) and (X_2, ρ_2), respectively, and that are (as unrooted trees) subspaces of a common metric space (Z, d_Z).

Lemma 4.27. For two rooted trees (X_1, d_{X_1}, ρ_1), and (X_2, d_{X_2}, ρ_2),

$$d_{\mathrm{GH^{root}}}((X_1, d_{X_1}, \rho_1), (X_2, d_{X_2}, \rho_2)) = \frac{1}{2} \inf_{\mathfrak{R}^{\mathrm{root}}} \mathrm{dis}(\mathfrak{R}^{\mathrm{root}}), \qquad (4.6)$$

where now the infimum is taken over all correspondences $\mathfrak{R}^{\mathrm{root}}$ between X_1 and X_2 with $(\rho_1, \rho_2) \in \mathfrak{R}^{\mathrm{root}}$.

Definition 4.28. Let (X_1, ρ_1) and (X_2, ρ_2) be two rooted compact $\mathbb{R}$-trees, and take $\varepsilon > 0$. A map f is called a root-invariant ε-isometry from (X_1, ρ_1) to (X_2, ρ_2) if $f(\rho_1) = \rho_2$, $\mathrm{dis}(f) < \varepsilon$ and $f(X_1)$ is an ε-net for X_2.

Lemma 4.29. Let (X_1, ρ_1) and (X_2, ρ_2) be two rooted compact $\mathbb{R}$-trees, and take $\varepsilon > 0$. Then the following hold.

(i) If $d_{\mathrm{GH^{root}}}((X_1, \rho_1), (X_2, \rho_2)) < \varepsilon$, then there exists a root-invariant 2ε-isometry from (X_1, ρ_1) to (X_2, ρ_2).

(ii)If there exists a root-invariant ε-isometry from (X_1, ρ_1) to (X_2, ρ_2), then

$$d_{\mathrm{GH^{root}}}((X_1, \rho_1), (X_2, \rho_2)) \leqslant \frac{3}{2}\varepsilon.$$

Proof. (i) Let $d_{\mathrm{GH^{root}}}((X_1, \rho_1), (X_2, \rho_2)) < \varepsilon$. By Lemma 4.27 there exists a correspondence $\mathfrak{R}^{\mathrm{root}}$ between X_1 and X_2 such that $(\rho_1, \rho_2) \in \mathfrak{R}^{\mathrm{root}}$ and $\mathrm{dis}(\mathfrak{R}^{\mathrm{root}}) < 2\varepsilon$.

Define $f : X_1 \to X_2$ by setting $f(\rho_1) = \rho_2$, and choosing $f(x)$ such that $(x, f(x)) \in \mathfrak{R}^{\mathrm{root}}$ for all $x \in X_1 \backslash \{\rho_1\}$.

Clearly, $\mathrm{dis}(f) \leqslant \mathrm{dis}(\mathfrak{R}^{\mathrm{root}}) < 2\varepsilon$.

To see that $f(X_1)$ is a 2ε-net for X_2, let $x_2 \in X_2$, and choose $x_1 \in X_1$ such that $(x_1, x_2) \in \mathfrak{R}^{\mathrm{root}}$. Then $d_{X_2}(f(x_1), x_2) \leqslant d_{X_1}(x_1, x_1) + \mathrm{dis}(\mathfrak{R}^{\mathrm{root}}) < 2\varepsilon$.

(ii) Let f be a root-invariant ε-isometry from (X_1, ρ_1) to (X_2, ρ_2). Define a correspondence $\mathfrak{R}_f^{\mathrm{root}} \subseteq X_1 \times X_2$ by

$$\mathfrak{R}_f^{\mathrm{root}} := \{(x_1, x_2) : d_{X_2}(x_2, f(x_1)) \leqslant \varepsilon\}. \tag{4.7}$$

Then $(\rho_1, \rho_2) \in \mathfrak{R}_f^{\mathrm{root}}$ and $\mathfrak{R}_f^{\mathrm{root}}$ is indeed a correspondence since $f(X_1)$ is a ε-net for X_2. If $(x_1, x_2), (y_1, y_2) \in \mathfrak{R}_f^{\mathrm{root}}$, then

$$
\begin{aligned}
|d_{X_1}(x_1, y_1) - d_{X_2}(x_2, y_2)| &\leqslant |d_{X_2}(f(x_1), f(y_1)) - d_{X_1}(x_1, y_1)| \\
&\quad + d_{X_2}(x_2, f(x_1)) + d_{X_2}(f(x_1), y_2) \\
&< 3\varepsilon.
\end{aligned}
\tag{4.8}
$$

Hence, $\mathrm{dis}(\mathfrak{R}_f^{\mathrm{root}}) < 3\varepsilon$ and, by (4.6),

$$d_{\mathrm{GH^{root}}}((X_1, \rho_1), (X_2, \rho_2)) \leqslant \frac{3}{2}\varepsilon.$$

□

We need the following compactness criterion, that is the analogue of Theorem 4.18 and can be proved the same way, noting that the analogue of Lemma 4.22 holds for $\mathbf{T}^{\mathrm{root}}$.

Lemma 4.30. *A subset $\mathcal{T} \subset \mathbf{T}^{\mathrm{root}}$ is relatively compact if and only if for every $\varepsilon > 0$ there exists a positive integer $N(\varepsilon)$ such that each $T \in \mathcal{T}$ has an ε-net with at most $N(\varepsilon)$ points.*

Theorem 4.31. *The metric space $(\mathbf{T}^{\mathrm{root}}, d_{GH^{\mathrm{root}}})$ is complete and separable.*

Proof. The proof follows very much the same lines as that of Theorem 4.23. The proof of separability is almost identical. The key step in establishing completeness is again to show that a Cauchy sequence in $\mathbf{T}^{\mathrm{root}}$ has a subsequential limit. This can be shown in the same manner as in the proof of Theorem 4.23, with an appeal to Lemma 4.30 replacing one to Theorem 4.18.   □

4.3.4 Rooted Subtrees and Trimming

A *rooted subtree* of a rooted $\mathbb{R}$-tree $(T, d, \rho) \in \mathbf{T}^{\mathrm{root}}$ is an element $(T^*, d^*, \rho^*) \in \mathbf{T}^{\mathrm{root}}$ that has a class representative that is a subspace of a class representative of (T, d, ρ), with the two roots coincident. Equivalently, any class representative of (T^*, d^*, ρ^*) can be isometrically embedded into any class representative of (T, d, ρ) via an isometry that maps roots to roots. We write $T^* \leq^{\mathrm{root}} T$ and note that $\leq^{\mathrm{root}}$ is a partial order on $\mathbf{T}^{\mathrm{root}}$.

For $\eta > 0$ define $R_\eta : \mathbf{T}^{\mathrm{root}} \to \mathbf{T}^{\mathrm{root}}$ to be the map that assigns to $(T, \rho) \in \mathbf{T}^{\mathrm{root}}$ the rooted subtree $(R_\eta(T), \rho)$ that consists of ρ and points $a \in T$ for which the subtree

$$S^{T,a} := \{x \in T : a \in [\rho, x]\}$$

(that is, the *subtree above* a) has height greater than or equal to η. Equivalently,

$$R_\eta(T) := \{x \in T : \exists y \in T \text{ such that } x \in [\rho, y],\ d_T(x, y) \geq \eta\} \cup \{\rho\}.$$

In particular, if T has height at most η, then $R_\eta(T)$ is just the trivial tree consisting of the root ρ. See Figure 4.6 for an example of this construction.

Lemma 4.32. *(i) The range of R_η consists of finite rooted trees (that is, rooted compact $\mathbb{R}$-trees with finitely many leaves).*
(ii) The map R_η is continuous.
(iii) The family of maps $(R_\eta)_{\eta > 0}$ is a semigroup; that is,

$$R_{\eta'} \circ R_{\eta''} = R_{\eta' + \eta''} \text{ for } \eta', \eta'' > 0.$$

In particular,

$$R_{\eta'}(T) \leq^{\mathrm{root}} R_{\eta''}(T) \text{ for } \eta' \geq \eta'' > 0.$$

(iv) For any $(T, \rho) \in \mathbf{T}^{\mathrm{root}}$,

$$d_{\mathrm{GH}^{\mathrm{root}}}((T, \rho), (R_\eta(T), \rho)) \leq d_{\mathrm{H}}(T, R_\eta(T)) \leq \eta,$$

where d_{H} is the Hausdorff metric on compact subsets of T induced by the metric ρ.

Lemma 4.33. *Consider a sequence $\{T_n\}_{n \in \mathbb{N}}$ of representatives of isometry classes of rooted compact trees in $(\mathbf{T}, d_{\mathrm{GH}^{\mathrm{root}}})$ with the following properties.*

- *Each set T_n is a subset of some common set U.*
- *Each tree T_n has the same root $\rho \in U$.*
- *The sequence $\{T_n\}_{n \in \mathbb{N}}$ is nondecreasing, that is, $T_1 \subseteq T_2 \subseteq \cdots \subseteq U$.*
- *Writing d_n for the metric on T_n, for $m < n$ the restriction of d_n to T_m coincides with d_m, so that there is a well-defined metric on $T := \bigcup_{n \in \mathbb{N}} T_n$ given by*

$$d(a, b) = d_n(a, b), \quad a, b \in T_n.$$

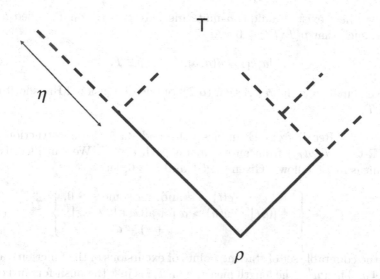

Fig. 4.6. Trimming a tree. The tree T consists of both the solid and dashed edges. The η-trimming $R_\eta(T)$ consists of the solid edges and is composed of the points of T that are distance at least η from some leaf of T.

- *The sequence of subsets $(T_n)_{n\in\mathbb{N}}$ is Cauchy in the Hausdorff distance with respect to d.*

Then the metric completion $\bar{T}$ of T is a compact $\mathbb{R}$-tree, and $d_{\mathrm{H}}(T_n, \bar{T}) \to 0$ as $n \to \infty$, where the Hausdorff distance is computed with respect to the extension of d to $\bar{T}$. In particular,

$$\lim_{n\to\infty} d_{\mathrm{GH^{root}}}((T_n, \rho), (\bar{T}, \rho)) = 0.$$

4.3.5 Length Measure on ℝ-Trees

Fix $(T, d, \rho) \in \mathbf{T}^{\mathrm{root}}$, and denote the Borel-$\sigma$-field on T by $\mathcal{B}(T)$. Write

$$T^o := \bigcup_{b\in T} [\rho, b[\tag{4.9}$$

for the *skeleton* of T.

Observe that if $T' \subset T$ is a dense countable set, then (4.9) holds with T replaced by T'. In particular, $T^o \in \mathcal{B}(T)$ and $\mathcal{B}(T)|_{T^o} = \sigma(\{]a, b[; a, b \in T'\})$, where

$$\mathcal{B}(T)\big|_{T^o} := \{A \cap T^o; \ A \in \mathcal{B}(T)\}$$

Hence, there exists a unique σ-finite measure $\mu = \mu^T$ on T, called *length measure*, such that $\mu(T \backslash T^o) = 0$ and

$$\mu(]a, b[) = d(a, b), \quad \forall\, a, b \in T. \tag{4.10}$$

In particular, μ is the restriction to T^o of one-dimensional Hausdorff measure on T.

Example 4.34. Recall from Examples 3.14 and 3.37 the construction of a rooted $\mathbb{R}$-tree (T_e, d_{T_e}) from an excursion path $e \in U$. We can identify the length measure as follows. Given $e \in U^\ell$ and $a \geqslant 0$, let

$$\mathcal{G}_a := \left\{ t \in [0, \ell] : \begin{array}{l} e(t) = a \text{ and, for some } \varepsilon > 0, \\ e(u) > a \text{ for all } u \in]t, t + \varepsilon[, \\ e(t + \varepsilon) = a. \end{array} \right\} \tag{4.11}$$

denote the countable set of starting points of excursions of the function e above the level a. Then μ^{T_e}, the length measure on T_e, is just the push-forward of the measure $\int_0^\infty da \sum_{t \in \mathcal{G}_a} \delta_t$ by the quotient map. Alternatively – see Figure 4.7 – write

$$\Gamma_e := \{(s, a) : \ s \in]0, \ell[, \ a \in [0, e(s)[\} \tag{4.12}$$

for the region between the time axis and the graph of e, and for $(s, a) \in \Gamma_e$ denote by $\underline{s}(e, s, a) := \sup\{r < s : e(r) = a\}$ and $\bar{s}(e, s, a) := \inf\{t > s : e(t) = a\}$ the start and finish of the excursion of e above level a that straddles time s. Then μ^{T_e} is the push-forward of the measure $\int_{\Gamma_e} ds \otimes da \frac{1}{\bar{s}(e,s,a) - \underline{s}(e,s,a)} \delta_{\underline{s}(e,s,a)}$ by the quotient map. We note that the measure μ^{T_e} appears in [1].

There is a simple recipe for the total length of a finite tree (that is, a tree with finitely many leaves).

Lemma 4.35. *Let* $(T, d, \rho) \in \mathbf{T}^{\mathrm{root}}$ *and suppose that* $\{x_0, \ldots, x_n\} \subset T$ *spans* T, *so that the root* ρ *and the leaves of* T *form a subset of* $\{x_0, \ldots, x_n\}$. *Then the total length of* T *(that is, the total mass of its length measure) is given by*

$$d(x_0, x_1) + \sum_{k=2}^n \bigwedge_{0 \leqslant i < j \leqslant k-1} \frac{1}{2} \left(d(x_k, x_i) + d(x_k, x_j) - d(x_i, x_j) \right)$$

$$= d(x_0, x_1) + \sum_{k=2}^n \bigwedge_{0 \leqslant i < j \leqslant k-1} (x_i \cdot x_j)_{x_k}$$

– see Figure 4.8.

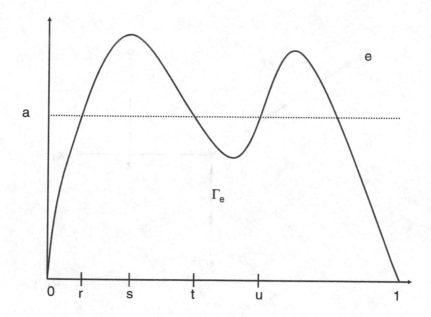

Fig. 4.7. Various objects associated with an excursion $e \in U^1$. The set of starting points of excursions of e above level a is $\mathcal{G}_a = \{r, u\}$. The region between the graph of e and the time axis is Γ_e. The start and finish of the excursion of e above level a that straddles time s are $\underline{s}(e, s, a) = r$ and $\bar{s}(e, s, a) = t$.

Proof. This follows from the observation that the distance from the point x_k to the segment $[x_i, x_j]$ is

$$\frac{1}{2}\left(d(x_k, x_i) + d(x_k, x_j) - d(x_i, x_j)\right) = (x_i \cdot x_j)_{x_k},$$

in the notation of Definition 3.7, and so the length of the segment connecting x_k, $2 \leqslant k \leqslant n$, to the subtree spanned by $x_0, \dots, x_{k-1}$ is

$$\bigwedge_{0 \leqslant i < j \leqslant k-1} \frac{1}{2}\left(d(x_k, x_i) + d(x_k, x_j) - d(x_i, x_j)\right).$$

□

The formula of Lemma 4.35 can be used to establish the following result, which implies that the function that sends a tree to its total length is lower semi-continuous (and, therefore, Borel). We refer the reader to Lemma 7.3 of [63] for the proof.

Lemma 4.36. *For $\eta > 0$, the map $T \mapsto \mu^T(R_\eta)$ (that is, the map that takes a tree to the total length of its η-trimming) is continuous.*

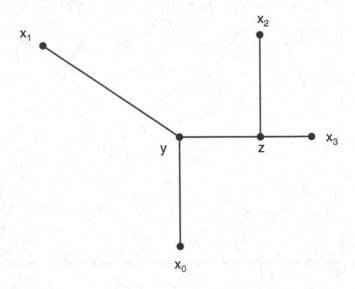

Fig. 4.8. The construction of Lemma 4.35. The total length of the tree is $d(x_0, x_1) + d(x_2, y) + d(x_3, z)$.

The following result, when combined with the compactness criterion Lemma 4.30, gives an alternative necessary and sufficient condition for a subset of $\mathbf{T}^{\mathrm{root}}$ to be relatively compact (Corollary 4.38 below).

Lemma 4.37. *Let $T \in \mathbf{T}^{\mathrm{root}}$ be such that $\mu^T(T) < \infty$. For each $\varepsilon > 0$ there is an ε-net for T of cardinality at most*

$$\left[\left(\frac{\varepsilon}{2} \right)^{-1} \mu^T(T) \right] \left[\left(\frac{\varepsilon}{2} \right)^{-1} \mu^T(T) + 1 \right].$$

Proof. Note that an $\frac{\varepsilon}{2}$-net for $R_{\frac{\varepsilon}{2}}(T)$ will be an ε-net for T. The set $T \backslash R_{\frac{\varepsilon}{2}}(T)^o$ is the union of a collection disjoint subtrees. Each leaf of $R_{\frac{\varepsilon}{2}}(T)$ belongs to a unique such subtree, and the diameter of each such subtree is at least $\frac{\varepsilon}{2}$. (There may also be other subtrees in the collection that don't contain leaves of $R_{\frac{\varepsilon}{2}}(T)$.) Thus, the number of leaves of $R_{\frac{\varepsilon}{2}}(T)$ is at most $\left(\frac{\varepsilon}{2} \right)^{-1} \mu^T(T)$. Enumerate the leaves of $R_{\frac{\varepsilon}{2}}(T)$ as $x_0, x_1, \ldots, x_n$. Each segment $[x_0, x_i]$, $1 \leqslant i \leqslant n$, of $R_{\frac{\varepsilon}{2}}(T)$ has an $\frac{\varepsilon}{2}$-net of cardinality at most $\left(\frac{\varepsilon}{2} \right)^{-1} d_T(x_0, x_i) + 1 \leqslant \left(\frac{\varepsilon}{2} \right)^{-1} \mu^T(T) + 1$. Therefore, by taking the union of these nets, $R_{\frac{\varepsilon}{2}}(T)$ has an $\frac{\varepsilon}{2}$-net of cardinality at most $\left[\left(\frac{\varepsilon}{2} \right)^{-1} \mu^T(T) \right] \left[\left(\frac{\varepsilon}{2} \right)^{-1} \mu^T(T) + 1 \right]$. $\square$

Corollary 4.38. *A subset* $\mathcal{T}$ *of* $(\mathbf{T}^{\mathrm{root}}, d_{\mathrm{GH^{root}}})$ *is relatively compact if and only if for all* $\varepsilon > 0$,

$$\sup\{\mu^T(R_\varepsilon(T)) : T \in \mathcal{T}\} < \infty.$$

Proof. The "only if" direction follows from continuity of $T \mapsto \mu^T(R_\varepsilon(T))$ obtained in Lemma 4.36.

Conversely, suppose that the condition of the corollary holds. Given $T \in \mathcal{T}$, an ε-net for $R_\varepsilon(T)$ is a 2ε-net for T. By Lemma 4.37, $R_\varepsilon(T)$ has an ε-net of cardinality at most

$$\left[\left(\frac{\varepsilon}{2}\right)^{-1} \mu^T(R_\varepsilon(T))\right]\left[\left(\frac{\varepsilon}{2}\right)^{-1} \mu^T(R_\varepsilon(T)) + 1\right].$$

By assumption, the last quantity is uniformly bounded in $T \in \mathcal{T}$. Hence, the set $\mathcal{T}$ is relatively compact by Lemma 4.30.   □

4.4 Weighted ℝ-Trees

A *weighted* ℝ-*tree* is a ℝ-tree (T, d) equipped with a probability measure ν on the Borel σ-field $\mathcal{B}(T)$. Write $\mathbf{T}^{\mathrm{wt}}$ for the space of weight-preserving isometry classes of weighted compact ℝ-trees, where we say that two weighted, compact ℝ-trees (X, d, ν) and (X', d', ν') are weight-preserving isometric if there exists an isometry ϕ between X and X' such that the *push-forward* of ν by ϕ is ν':

$$\nu' = \phi_*\nu := \nu \circ \phi^{-1}. \tag{4.13}$$

It is clear that the property of being weight-preserving isometric is an equivalence relation.

Example 4.39. Recall from Examples 3.14 and 3.37 the construction of a compact ℝ-tree from an excursion path $e \in U^\ell$. Such a ℝ-tree has a canonical weight, namely, the push-forward of normalized Lebesgue measure on $[0, \ell]$ by the quotient map that appears in the construction.

We want to equip $\mathbf{T}^{\mathrm{wt}}$ with a Gromov-Hausdorff type of distance that incorporates the weights on the trees.

Lemma 4.40. *Let* (X, d_X) *and* (Y, d_Y) *be two compact real trees such that* $d_{\mathrm{GH}}((X, d_X), (Y, d_Y)) < \varepsilon$ *for some* $\varepsilon > 0$. *Then there exists a measurable* 3ε-*isometry from* X *to* Y.

Proof. If $d_{\mathrm{GH}}((X, d_X), (Y, d_Y)) < \varepsilon$, then by Theorem 4.11 there exists a correspondence $\Re$ between X and Y such that $\mathrm{dis}(\Re) < 2\varepsilon$. Since (X, d_X) is compact there exists a finite ε-net in X. We claim that for each such finite ε-net $S^{X,\varepsilon} = \{x_1, ..., x_{N^\varepsilon}\} \subseteq X$, any set $S^{Y,\varepsilon} = \{y_1, ..., y_{N^\varepsilon}\} \subseteq Y$ such that

$(x_i, y_i) \in \Re$ for all $i \in \{1, 2, ..., N^\varepsilon\}$ is a 3ε-net in Y. To see this, fix $y \in Y$. We have to show the existence of $i \in \{1, 2, ..., N^\varepsilon\}$ with $d_Y(y_i, y) < 3\varepsilon$. For that choose $x \in X$ such that $(x, y) \in \Re$. Since $S^{X,\varepsilon}$ is an ε-net in X there exists an $i \in \{1, 2, ..., N^\varepsilon\}$ such that $d_X(x_i, x) < \varepsilon$. $(x_i, y_i) \in \Re$ implies, therefore, that $|d_X(x_i, x) - d_Y(y_i, y)| \leqslant \operatorname{dis}(\Re) < 2\varepsilon$. Hence, $d_Y(y_i, y) < 3\varepsilon$.

Furthermore, we may decompose X into N^ε possibly empty measurable disjoint subsets of X by letting $X^{1,\varepsilon} := \mathcal{B}(x_1, \varepsilon)$, $X^{2,\varepsilon} := \mathcal{B}(x_2, \varepsilon) \backslash X^{1,\varepsilon}$, and so on, where $\mathcal{B}(x, r)$ is the open ball $\{x' \in X : d_X(x, x') < r\}$. Then f defined by $f(x) = y_i$ for $x \in X^{i,\varepsilon}$ is obviously a measurable 3ε-isometry from X to Y. □

We also need to recall the definition of the Prohorov distance between two probability measures – see, for example, [57]. Given two probability measures μ and ν on a metric space (X, d) with the corresponding collection of closed sets denoted by $\mathcal{C}$, the Prohorov distance between them is

$$d_{\mathrm{P}}(\mu, \nu) := \inf\{\varepsilon > 0 : \mu(C) \leqslant \nu(C^\varepsilon) + \varepsilon \text{ for all } C \in \mathcal{C}\},$$

where $C^\varepsilon := \{x \in X : \inf_{y \in C} d(x, y) < \varepsilon\}$. The Prohorov distance is a metric on the collection of probability measures on X. The following result shows that if we push measures forward with a map having a small distortion, then Prohorov distances can't increase too much.

Lemma 4.41. *Suppose that (X, d_X) and (Y, d_Y) are two metric spaces, $f : X \to Y$ is a measurable map with $\operatorname{dis}(f) \leqslant \varepsilon$, and μ and ν are two probability measures on X. Then*

$$d_{\mathrm{P}}(f_* \mu, f_* \nu) \leqslant d_{\mathrm{P}}(\mu, \nu) + \varepsilon.$$

Proof. Suppose that $d_{\mathrm{P}}(\mu, \nu) < \delta$. By definition, $\mu(C) \leqslant \nu(C^\delta) + \delta$ for all closed sets $C \in \mathcal{C}$. If D is a closed subset of Y, then

$$\begin{aligned}
f_* \mu(D) &= \mu(f^{-1}(D)) \\
&\leqslant \mu(\overline{f^{-1}(D)}) \\
&\leqslant \nu(\overline{f^{-1}(D)}^\delta) + \delta \\
&= \nu(f^{-1}(D)^\delta) + \delta.
\end{aligned}$$

Now $x' \in f^{-1}(D)^\delta$ means there is $x'' \in X$ such that $d_X(x', x'') < \delta$ and $f(x'') \in D$. By the assumption that $\operatorname{dis}(f) \leqslant \varepsilon$, we have $d_Y(f(x'), f(x'')) < \delta + \varepsilon$. Hence, $f(x') \in D^{\delta+\varepsilon}$. Thus,

$$f^{-1}(D)^\delta \subseteq f^{-1}(D^{\delta+\varepsilon})$$

and we have

$$f_* \mu(D) \leqslant \nu(f^{-1}(D^{\delta+\varepsilon})) + \delta = f_* \nu(D^{\delta+\varepsilon}) + \delta,$$

so that $d_{\mathrm{P}}(f_* \mu, f_* \nu) \leqslant \delta + \varepsilon$, as required. □

We are now in a position to define the weighted Gromov-Hausdorff distance between the two compact, weighted ℝ-trees (X, d_X, ν_X) and (Y, d_Y, ν_Y). For $\varepsilon > 0$, set

$$F_{X,Y}^\varepsilon := \{\text{measurable } \varepsilon\text{-isometries from } X \text{ to } Y\}. \qquad (4.14)$$

Put

$$\Delta_{\mathrm{GH^{wt}}}(X, Y)$$

$$:= \inf \left\{ \varepsilon > 0 : \begin{array}{c} \text{there exist } f \in F_{X,Y}^\varepsilon, g \in F_{Y,X}^\varepsilon \text{ such that} \\ d_{\mathrm{P}}(f_*\nu_X, \nu_Y) \leqslant \varepsilon, \, d_{\mathrm{P}}(\nu_X, g_*\nu_Y) \leqslant \varepsilon \end{array} \right\}. \qquad (4.15)$$

Note that the set on the right hand side is non-empty because X and Y are compact, and, therefore, bounded. It will turn out that $\Delta_{\mathrm{GH^{wt}}}$ satisfies all the properties of a metric except the triangle inequality. To rectify this, let

$$d_{\mathrm{GH^{wt}}}(X, Y) := \inf \left\{ \sum_{i=1}^{n-1} \Delta_{\mathrm{GH^{wt}}}(Z_i, Z_{i+1})^{\frac{1}{4}} \right\}, \qquad (4.16)$$

where the infimum is taken over all finite sequences of compact, weighted ℝ-trees $Z_1, \ldots Z_n$ with $Z_1 = X$ and $Z_n = Y$.

Lemma 4.42. *The map* $d_{\mathrm{GH^{wt}}} : \mathbf{T}^{\mathrm{wt}} \times \mathbf{T}^{\mathrm{wt}} \to \mathbb{R}_+$ *is a metric on* $\mathbf{T}^{\mathrm{wt}}$. *Moreover,*

$$\frac{1}{2} \Delta_{\mathrm{GH^{wt}}}(X, Y)^{\frac{1}{4}} \leqslant d_{\mathrm{GH^{wt}}}(X, Y) \leqslant \Delta_{\mathrm{GH^{wt}}}(X, Y)^{\frac{1}{4}}$$

for all $X, Y \in \mathbf{T}^{\mathrm{wt}}$.

Proof. It is immediate from (4.15) that the map $\Delta_{\mathrm{GH^{wt}}}$ is symmetric.
    We next claim that

$$\Delta_{\mathrm{GH^{wt}}}\big((X, d_X, \nu_X), (Y, d_Y, \nu_Y)\big) = 0, \qquad (4.17)$$

if and only if (X, d_X, ν_X) and (Y, d_Y, ν_Y) are weight-preserving isometric. The "if" direction is immediate. Note first for the converse that (4.17) implies that for all $\varepsilon > 0$ there exists an ε-isometry from X to Y, and, therefore, by Corollary 4.13, $d_{\mathrm{GH}}\big((X, d_X), (Y, d_Y)\big) < 2\varepsilon$. Thus, $d_{\mathrm{GH}}\big((X, d_X), (Y, d_Y)\big) = 0$, and it follows from Theorem 4.14 that (X, d_X) and (Y, d_Y) are isometric. Checking the proof of that result, we see that we can construct an isometry $f : X \to Y$ by taking any dense countable set $S \subset X$, any sequence of functions (f_n) such that f_n is an ε_n-isometry with $\varepsilon_n \to 0$ as $n \to \infty$, and letting f be $\lim_k f_{n_k}$ along any subsequence such that the limit exists for all $x \in S$ (such a subsequence exists by the compactness of Y). Therefore, fix some dense subset $S \subset X$ and suppose without loss of generality that we have an isometry $f : X \to Y$ given by $f(x) = \lim_{n\to\infty} f_n(x)$, $x \in S$, where $f_n \in F_{X,Y}^{\varepsilon_n}$, $d_{\mathrm{P}}(f_{n*}\nu_X, \nu_Y) \leqslant \varepsilon_n$, and $\lim_{n\to\infty} \varepsilon_n = 0$. We will be done if we can show that $f_* \nu_X = \nu_Y$. If μ_X is a discrete measure with atoms belonging to S, then

$$\begin{aligned} d_\mathrm{P}(f_* \nu_X, \nu_Y) &\leqslant \limsup_n \Big[d_\mathrm{P}(f_{n*} \nu_X, \nu_Y) + d_\mathrm{P}(f_{n*} \mu_X, f_{n*} \nu_X) \\ &\qquad + d_\mathrm{P}(f_* \mu_X, f_{n*} \mu_X) + d_\mathrm{P}(f_* \nu_X, f_* \mu_X) \Big] \\ &\leqslant 2 d_\mathrm{P}(\mu_X, \nu_X), \end{aligned} \tag{4.18}$$

where we have used Lemma 4.41 and the fact that $\lim_{n \to \infty} d_\mathrm{P}(f_* \mu_X, f_{n*} \mu_X) = 0$ because of the pointwise convergence of f_n to f on S. Because we can choose μ_X so that $d_\mathrm{P}(\mu_X, \nu_X)$ is arbitrarily small, we see that $f_* \nu_X = \nu_Y$, as required.

Now consider three spaces (X, d_X, ν_X), (Y, d_Y, ν_Y), and (Z, d_Z, ν_Z) in $\mathbf{T}^{\mathrm{wt}}$, and constants $\varepsilon, \delta > 0$, such that $\Delta_{\mathrm{GH^{wt}}}\big((X, d_X, \nu_X), (Y, d_Y, \nu_Y)\big) < \varepsilon$ and $\Delta_{\mathrm{GH^{wt}}}\big((Y, d_Y, \nu_Y), (Z, d_Z, \nu_Z)\big) < \delta$. Then there exist $f \in F_{X,Y}^\varepsilon$ and $g \in F_{Y,Z}^\delta$ such that $d_\mathrm{P}(f_* \nu_X, \nu_Y) < \varepsilon$ and $d_\mathrm{P}(g_* \nu_Y, \nu_Z) < \delta$. Note that $g \circ f \in F_{X,Z}^{\varepsilon + \delta}$. Moreover, by Lemma 4.41

$$d_\mathrm{P}((g \circ f)_* \nu_X, \nu_Z) \leqslant d_\mathrm{P}(g_* \nu_Y, \nu_Z) + d_\mathrm{P}(g_* f_* \nu_X, g_* \nu_Y) < \delta + \varepsilon + \delta. \tag{4.19}$$

This, and a similar argument with the roles of X and Z interchanged, shows that

$$\Delta_{\mathrm{GH^{wt}}}(X, Z) \leqslant 2 \left[\Delta_{\mathrm{GH^{wt}}}(X, Y) + \Delta_{\mathrm{GH^{wt}}}(Y, Z) \right]. \tag{4.20}$$

The second inequality in the statement of the lemma is clear. In order to see the first inequality, it suffices to show that for any $Z_1, \dots Z_n$ we have

$$\Delta_{\mathrm{GH^{wt}}}(Z_1, Z_n)^{\frac{1}{4}} \leqslant 2 \sum_{i=1}^{n-1} \Delta_{\mathrm{GH^{wt}}}(Z_i, Z_{i+1})^{\frac{1}{4}}. \tag{4.21}$$

We will establish (4.21) by induction. The inequality certainly holds when $n = 2$. Suppose it holds for $2, \dots, n - 1$. Write S for the value of the sum on the right hand side of (4.21). Put

$$k := \max \left\{ 1 \leqslant m \leqslant n - 1 : \sum_{i=1}^{m-1} \Delta_{\mathrm{GH^{wt}}}(Z_i, Z_{i+1})^{\frac{1}{4}} \leqslant S/2 \right\}. \tag{4.22}$$

By the inductive hypothesis and the definition of k,

$$\Delta_{\mathrm{GH^{wt}}}(Z_1, Z_k)^{\frac{1}{4}} \leqslant 2 \sum_{i=1}^{k-1} \Delta_{\mathrm{GH^{wt}}}(Z_i, Z_{i+1})^{\frac{1}{4}} \leqslant 2(S/2) = S. \tag{4.23}$$

Of course,

$$\Delta_{\mathrm{GH^{wt}}}(Z_k, Z_{k+1})^{\frac{1}{4}} \leqslant S \tag{4.24}$$

By definition of k,

$$\sum_{i=1}^{k} \Delta_{\mathrm{GH^{wt}}}(Z_i, Z_{i+1})^{\frac{1}{4}} > S/2,$$

so that once more by the inductive hypothesis,

$$
\begin{aligned}
\Delta_{\mathrm{GH^{wt}}}(Z_{k+1}, Z_n)^{\frac{1}{4}} &\leqslant 2 \sum_{i=k+1}^{n-1} \Delta_{\mathrm{GH^{wt}}}(Z_i, Z_{i+1})^{\frac{1}{4}} \\
&= 2S - 2 \sum_{i=1}^{k} \Delta_{\mathrm{GH^{wt}}}(Z_i, Z_{i+1})^{\frac{1}{4}} \\
&\leqslant S.
\end{aligned}
\tag{4.25}
$$

From (4.23), (4.24), (4.25) and two applications of (4.20) we have

$$
\begin{aligned}
\Delta_{\mathrm{GH^{wt}}}(Z_1, Z_n)^{\frac{1}{4}} &\leqslant \{4[\Delta_{\mathrm{GH^{wt}}}(Z_1, Z_k) + \Delta_{\mathrm{GH^{wt}}}(Z_k, Z_{k+1}) \\
&\quad + \Delta_{\mathrm{GH^{wt}}}(Z_{k+1}, Z_n)]\}^{\frac{1}{4}} \\
&\leqslant (4 \times 3 \times S^4)^{\frac{1}{4}} \\
&\leqslant 2S,
\end{aligned}
\tag{4.26}
$$

as required.

It is obvious by construction that $d_{\mathrm{GH^{wt}}}$ satisfies the triangle inequality. The other properties of a metric follow from the corresponding properties we have already established for $\Delta_{\mathrm{GH^{wt}}}$ and the bounds in the statement of the lemma that we have already established.   □

The procedure we used to construct the weighted Gromov-Hausdorff metric $d_{\mathrm{GH^{wt}}}$ from the semi-metric $\Delta_{\mathrm{GH^{wt}}}$ was adapted from a proof in [88] of the celebrated result of Alexandroff and Urysohn on the metrizability of uniform spaces. That proof was, in turn, adapted from earlier work of Frink and Bourbaki. The choice of the power $\frac{1}{4}$ is not particularly special, any sufficiently small power would have worked.

Proposition 4.43. *A subset* **D** *of* $(\mathbf{T}^{\mathrm{wt}}, d_{\mathrm{GH^{wt}}})$ *is relatively compact if and only if the subset* $\mathbf{E} := \{(T, d) : (T, d, \nu) \in \mathbf{D}\}$ *in* $(\mathbf{T}, d_{\mathrm{GH}})$ *is relatively compact.*

Proof. The "only if" direction is clear. Assume for the converse that **E** is relatively compact. Suppose that $((T_n, d_{T_n}, \nu_{T_n}))_{n \in \mathbb{N}}$ is a sequence in **D**. By assumption, $((T_n, d_{T_n}))_{n \in \mathbb{N}}$ has a subsequence converging to some point (T, d_T) of $(\mathbf{T}, d_{\mathrm{GH}})$. For ease of notation, we will renumber and also denote this subsequence by $((T_n, d_{T_n}))_{n \in \mathbb{N}}$. For brevity, we will also omit specific mention of the metric on a real tree when it is clear from the context.

By Proposition 4.15, for each $\varepsilon > 0$ there is a finite ε-net T^ε in T and for each $n \in \mathbb{N}$ a finite ε-net $T_n^\varepsilon := \{x_n^{\varepsilon,1}, ..., x_n^{\varepsilon, \#T_n^\varepsilon}\}$ in T_n such that

$d_{\mathrm{GH}}(T_n^\varepsilon, T^\varepsilon) \to 0$ as $n \to \infty$. Without loss of generality, we may assume that $\#T_n^\varepsilon = \#T^\varepsilon$ for all $n \in \mathbb{N}$. We may begin with the balls of radius ε around each point of $\#T_n^\varepsilon$ and decompose T_n into $\#T_n^\varepsilon$ possibly empty, disjoint, measurable sets $\{T_n^{\varepsilon,1}, ..., T_n^{\varepsilon,\#T^\varepsilon}\}$ of radius no greater than ε. Define a measurable map $f_n : T_n \to T_n^\varepsilon$ by $f_n^\varepsilon(x) = x_n^{\varepsilon,i}$ if $x \in T_n^{\varepsilon,i}$ and let g_n^ε be the inclusion map from T_n^ε to T_n. By construction, f_n^ε and g_n^ε are ε-isometries. Moreover, $d_{\mathrm{P}}\big((g_n^\varepsilon)_*(f_n^\varepsilon)_*\nu_n, \nu_n\big) < \varepsilon$ and, of course, $d_{\mathrm{P}}\big((f_n^\varepsilon)_*\nu_n, (f_n^\varepsilon)_*\nu_n\big) = 0$. Thus, $\Delta_{\mathrm{GH^{wt}}}\big((T_n^\varepsilon, (f_n^\varepsilon)_*\nu_n), (T_n, \nu_n)\big) \leqslant \varepsilon$. By similar reasoning, if we define $h_n^\varepsilon : T_n^\varepsilon \to T^\varepsilon$ by $x_n^{\varepsilon,i} \mapsto x^{\varepsilon,i}$, then

$$\Delta_{\mathrm{GH^{wt}}}\big((T_n^\varepsilon, (f_n^\varepsilon)_*\nu_n), (T^\varepsilon, (h_n^\varepsilon)_*\nu_n)\big) \to 0$$

as $n \to \infty$. Since T^ε is finite, by passing to a subsequence (and relabeling as before) we have $\lim_{n\to\infty} d_{\mathrm{P}}\big((h_n^\varepsilon)_*\nu_n, \nu^\varepsilon\big) = 0$ for some probability measure ν^ε on T^ε. Hence,

$$\lim_{n\to\infty} \Delta_{\mathrm{GH^{wt}}}\big((T^\varepsilon, (h_n^\varepsilon)_*\nu_n), (T^\varepsilon, \nu^\varepsilon)\big) = 0.$$

Therefore, by Lemma 4.42,

$$\limsup_{n\to\infty} d_{\mathrm{GH^{wt}}}\big((T_n, \nu_n), (T^\varepsilon, (h_n^\varepsilon)_*\nu_n)\big) \leqslant \varepsilon^{\frac{1}{4}}.$$

Now, since (T, d_T) is compact, the family of measures $\{\nu^\varepsilon : \varepsilon > 0\}$ is relatively compact, and so there is a probability measure ν on T such that ν^ε converges to ν in the Prohorov distance along a subsequence $\varepsilon \downarrow 0$. Hence, by arguments similar to the above, along the same subsequence $\Delta_{\mathrm{GH^{wt}}}((T^\varepsilon, \nu^\varepsilon), (T, \nu))$ converges to 0. Again applying Lemma 4.42, we have that $d_{\mathrm{GH^{wt}}}((T^\varepsilon, \nu^\varepsilon), (T, \nu))$ converges to 0 along this subsequence.

Combining the foregoing, we see that by passing to a suitable subsequence and relabeling, $d_{\mathrm{GH^{wt}}}((T_n, \nu_n), (T, \nu))$ converges to 0, as required. □

Theorem 4.44. *The metric space* $(\mathbf{T}^{\mathrm{wt}}, d_{\mathrm{GH^{wt}}})$ *is complete and separable.*

Proof. Separability follows readily from the separability of $(\mathbf{T}, d_{\mathrm{GH}})$ and the separability with respect to the Prohorov distance of the probability measures on a fixed complete, separable metric space – see, for example, [57]) – and Lemma 4.42.

It remains to establish completeness. By a standard argument, it suffices to show that any Cauchy sequence in $\mathbf{T}^{\mathrm{wt}}$ has a convergent subsequence. Let $(T_n, d_{T_n}, \nu_n)_{n\in\mathbb{N}}$ be a Cauchy sequence in $\mathbf{T}^{\mathrm{wt}}$. Then $(T_n, d_{T_n})_{n\in\mathbb{N}}$ is a Cauchy sequence in $\mathbf{T}$ by Lemma 4.42. By Theorem 1 in [63] there is a $T \in \mathbf{T}$ such that $d_{\mathrm{GH}}(T_n, T) \to 0$, as $n \to \infty$. In particular, the sequence $(T_n, d_{T_n})_{n\in\mathbb{N}}$ is relatively compact in $\mathbf{T}$, and, therefore, by Proposition 4.43, $(T_n, d_{T_n}, \nu_n)_{n\in\mathbb{N}}$ is relatively compact in $\mathbf{T}^{\mathrm{wt}}$. Thus, $(T_n, d_{T_n})_{n\in\mathbb{N}}$ has a convergent subsequence, as required. □

5

Root Growth with Re-Grafting

5.1 Background and Motivation

Recall the special case of the tree-valued Markov chain that was used in the proof of the Markov chain tree theorem, Theorem 2.1, when the underlying Markov chain is the process on $\{1, 2, \ldots, n\}$ that picks a new state uniformly at each stage.

Algorithm 5.1.

- *Start with a rooted (combinatorial) tree on n labeled vertices $\{1, 2, \ldots, n\}$.*
- *Pick a vertex v uniformly from*
  $\{1, 2, \ldots, n\} \backslash \{$ current root $\}$.
- *Erase the edge leading from v towards the current root.*
- *Insert an edge from the current root to v and make v the new root.*
- *Repeat.*

We know that this chain converges in distribution to the uniform distribution on rooted trees with n labeled vertices.

Imagine that we do the following.

- Start with a rooted subtree (that is, one with the same root as the "big" tree).
- At each step of the chain, update the subtree by removing and adding edges as they are removed and added in the big tree and adjoining the new root of the big tree to the subtree if it isn't in the current subtree.

The subtree will evolve via two mechanisms that we might call *root growth* and *re-grafting*. Root growth occurs when the new root isn't in the current subtree, and so the new tree has an extra vertex, the new root, that is connected to the old root by a new edge. Re-grafting occurs when the new root is in the current subtree: it has the effect of severing the edge leading to a subtree of the current subtree and re-attaching it to the current root by a new edge. See Figure 5.1.

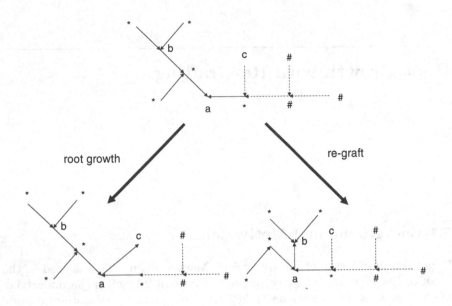

Fig. 5.1. Root growth and re-graft moves. The big tree with $n = 11$ vertices consists of the solid and dashed edges in all three diagrams. In the top diagram, the current subtree has the solid edges and the vertices marked a, b, $*$. The vertices marked c and $\#$ are in the big tree but not the current subtree. The big tree and the current subtree are rooted at a. The bottom left diagram shows the result of a root growth move: the vertex c now belongs to the new subtree, it is the root of the new big tree and the new subtree, and is connected to the old root a by an edge. The vertices marked $\#$ are not in the new subtree. The bottom right diagram shows the result of a re-graft move: the vertex b is the root of the new big tree and the new subtree, and it is connected to the old root a by an edge. The vertices marked c and $\#$ are not in the new subtree.

Now consider what happens as n becomes large and we follow a rooted subtree that originally has $\approx \sqrt{n}$ vertices. Replace edges of length 1 with edges of length $\frac{1}{\sqrt{n}}$ and speed up time by $\sqrt{n}$.

In the limit as $n \to \infty$, it seems reasonable that we have a $\mathbb{R}$-tree-valued process with the following *root growth with re-grafting* dynamics.

- The edge leading to the root of the evolving tree grows at unit speed.
- Cuts rain down on the tree at unit rate per length$\times$time, and the subtree above each cut is pruned off and re-attached at the root.

We will establish a closely related result in Section 5.4. Namely, we will show that if we have a sequence of chains following the dynamics of

Algorithm 5.1 such that the initial combinatorial tree of the nth chain re-scaled by $\sqrt{n}$ converges in the Gromov–Hausdorff distance to some compact $\mathbb{R}$-tree, then if we re-scale space and time by $\sqrt{n}$ in the nth chain we get weak convergence to a process with the root growth with re-grafting dynamics.

This latter result might seem counter-intuitive, because now we are working with the whole tree with n vertices rather than a subtree with $\approx \sqrt{n}$ vertices. However, the assumption that the initial condition scaled by $\sqrt{n}$ converges to some compact $\mathbb{R}$-tree means that asymptotically most vertices are close to the leaves and re-arranging the subtrees above such vertices has a negligible effect in the limit.

Before we can establish such a convergence result, we need to show that the root growth with re-grafting dynamics make sense even for compact trees with infinite total length. Such trees are the sort that will typically arise in the limit when we re-scale trees with n vertices by $\sqrt{n}$. This is not a trivial matter, as the set of times at which cuts appear will be dense and so the intuitive description of the dynamics does not make rigorous sense. See Theorem 5.5 for the details.

Given that the chain of Algorithm 5.1 converges at large times to the uniform rooted tree on n labeled vertices and that the uniform tree on n labeled vertices converges after suitable re-scaling to the Brownian continuum random tree as $n \to \infty$, it seems reasonable that the root growth with re-grafting process should converge at large times to the Brownian continuum random tree and that the Brownian continuum random tree should be the unique stationary distribution. We establish that this is indeed the case in Section 5.3. An important ingredient in the proofs of these facts will be Proposition 5.7, which says that the root growth with re-grafting process started from the trivial tree consisting of a single point is related to the Poisson line-breaking construction of the Brownian continuum random tree in Section 2.5 in the same manner that the chain of Algorithm 5.1 is related to Algorithm 2.4 for generating uniform rooted labeled trees. This is, of course, what we should expect, because the Poisson line-breaking construction arises as a limit of Algorithm 2.4 when the number of vertices goes to infinity.

5.2 Construction of the Root Growth with Re-Grafting Process

5.2.1 Outline of the Construction

- We want to construct a $\mathbf{T}^{\text{root}}$-valued process X with the root growth and re-grafting dynamics.
- Fix $(T, d, \rho) \in \mathbf{T}^{\text{root}}$. This will be X_0.
- We will construct simultaneously for each finite rooted subtree $T^* \leq^{\text{root}} T$ a process X^{T^*} with $X_0^{T^*} = T^*$ that evolves according to the root growth with re-grafting dynamics.

- We will carry out this construction in such a way that if T^* and T^{**} are two finite subtrees with $T^* \leq^{\text{root}} T^{**}$, then $X_t^{T^*} \leq^{\text{root}} X_t^{T^{**}}$ and the cut points for X^{T^*} are those for $X^{T^{**}}$ that happen to fall on $X_{\tau-}^{T^*}$ for a corresponding cut time τ of $X^{T^{**}}$. Cut times τ for $X^{T^{**}}$ for which the corresponding cut point does not fall on $X_{\tau-}^{T^*}$ are not cut times for X^{T^*}.

- The tree (T, ρ) is a rooted Gromov–Hausdorff limit of finite $\mathbb{R}$-trees with root ρ (indeed, any subtree of (T, ρ) that is spanned by the union of a finite ε-net and $\{\rho\}$ is a finite $\mathbb{R}$-tree that has rooted Gromov–Hausdorff distance less than ε from (T, ρ)).
  In particular, (T, ρ) is the "smallest" rooted compact $\mathbb{R}$-tree that contains all of the finite rooted subtrees of (T, ρ).

- Because of the consistent projective nature of the construction, we can define $X_t := X_t^T$ for $t \geq 0$ as the "smallest" element of $\mathbf{T}^{\text{root}}$ that contains $X_t^{T^*}$, for all finite trees $T^* \leq^{\text{root}} T$.

5.2.2 A Deterministic Construction

It will be convenient to work initially in a setting where the cut times and cut points are fixed.

There are two types of cut points: those that occur at points that were present in the initial tree T and those that occur at points that were added due to subsequent root growth.

Accordingly, we consider two countable subsets $\pi_0 \subset \mathbb{R}^{++} \times T^o$ and $\pi \subset \{(t, x) \in \mathbb{R}^{++} \times \mathbb{R}^{++} : x \leq t\}$. See Figure 5.2.

Assumption 5.2. *Suppose that the sets π_0 and π have the following properties.*

(a) For all $t_0 > 0$, each of the sets $\pi_0 \cap (\{t_0\} \times T^o)$ and $\pi \cap (\{t_0\} \times]0, t_0])$ has at most one point and at least one of these sets is empty.

(b) For all $t_0 > 0$ and all finite subtrees $T' \subseteq T$, the set $\pi_0 \cap (]0, t_0] \times T')$ is finite.

(c) For all $t_0 > 0$, the set $\pi \cap \{(t, x) \in \mathbb{R}^{++} \times \mathbb{R}^{++} : x \leq t \leq t_0\}$ is finite.

Remark 5.3. Conditions (a)–(c) of Assumption 5.2 will hold almost surely if π_0 and π are realizations of Poisson point processes with respective intensities $\lambda \otimes \mu$ and $\lambda \otimes \lambda$ (where λ is Lebesgue measure), and it is this random mechanism that we will introduce later to produce a stochastic process having the root growth with re-grafting dynamics.

Consider a finite rooted subtree $T^* \leq^{\text{root}} T$. It will avoid annoying circumlocutions about equivalence via root-invariant isometries if we work with particular class representatives for T^* and T, and, moreover, suppose that T^* is embedded in T.

Put $\tau_0^* := 0$, and let $0 < \tau_1^* < \tau_2^* < \ldots$ (the cut times for X^{T^*}) be the points of $\{t > 0 : \pi_0(\{t\} \times T^*) > 0\} \cup \{t > 0 : \pi(\{t\} \times \mathbb{R}^{++}) > 0\}$.

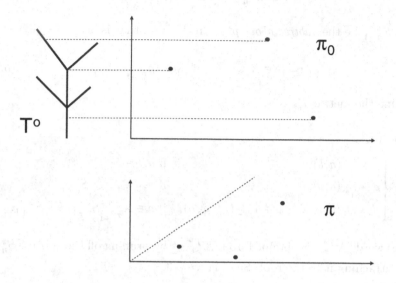

Fig. 5.2. The sets of points π_0 and π

Step 1 (Root growth). At any time $t \geqslant 0$, $X_t^{T^*}$ as a set is given by the disjoint union $T^* \sqcup]0, t]$. For $t > 0$, the root of $X_t^{T^*}$ is the point $\rho_t := t \in]0, t]$. The metric $d_t^{T^*}$ on $X_t^{T^*}$ is defined inductively as follows.

Set $d_0^{T^*}$ to be the metric on $X_0^{T^*} = T^*$; that is, $d_0^{T^*}$ is the restriction of d to T^*. Suppose that $d_t^{T^*}$ has been defined for $0 \leqslant t \leqslant \tau_n^*$. Define $d_t^{T^*}$ for $\tau_n^* < t < \tau_{n+1}^*$ by

$$d_t^{T^*}(a, b) := \begin{cases} d_{\tau_n^*}(a, b), & \text{if } a, b \in X_{\tau_n^*}^{T^*}, \\ |b - a|, & \text{if } a, b \in]\tau_n^*, t], \\ |a - \tau_n^*| + d_{\tau_n^*}(\rho_{\tau_n^*}, b), & \text{if } a \in]\tau_n^*, t], b \in X_{\tau_n^*}^{T^*}. \end{cases} \quad (5.1)$$

Step 2 (Re-Grafting). Note that the left-limit $X_{\tau_{n+1}^*-}^{T^*}$ exists in the rooted Gromov–Hausdorff metric. As a set this left-limit is the disjoint union

$$X_{\tau_n^*}^{T^*} \sqcup]\tau_n^*, \tau_{n+1}^*] = T^* \sqcup]0, \tau_{n+1}^*],$$

and the corresponding metric $d_{\tau_{n+1}^*-}$ is given by a prescription similar to (5.1).

Define the $(n+1)^{\text{st}}$ cut point for X^{T^*} by

$$p_{n+1}^* := \begin{cases} a \in T^*, & \text{if } \pi_0(\{(\tau_{n+1}^*, a)\}) > 0, \\ x \in]0, \tau_{n+1}^*], & \text{if } \pi(\{(\tau_{n+1}^*, x)\}) > 0. \end{cases}$$

Let S_{n+1}^* be the *subtree above* p_{n+1}^* in $X_{\tau_{n+1}^*-}^{T^*}$, that is,

$$S_{n+1}^* := \{b \in X_{\tau_{n+1}^*-}^{T^*} : p_{n+1}^* \in [\rho_{\tau_{n+1}^*-}, b[\}. \tag{5.2}$$

Define the metric $d_{\tau_{n+1}^*}$ by

$$d_{\tau_{n+1}^*}(a, b)$$
$$:= \begin{cases} d_{\tau_{n+1}^*-}(a, b), & \text{if } a, b \in S_{n+1}^*, \\ d_{\tau_{n+1}^*-}(a, b), & \text{if } a, b \in X_{\tau_{n+1}^*}^{T^*} \backslash S_{n+1}^*, \\ d_{\tau_{n+1}^*-}(a, \rho_{\tau_{n+1}^*}) + d_{\tau_{n+1}^*-}(p_{n+1}^*, b), & \text{if } a \in X_{\tau_{n+1}^*}^{T^*} \backslash S_{n+1}^*, b \in S_{n+1}^*. \end{cases}$$

In other words $X_{\tau_{n+1}^*}^{T^*}$ is obtained from $X_{\tau_{n+1}^*-}^{T^*}$ by pruning off the subtree S_{n+1}^* and re-attaching it to the root. See Figure 5.3.

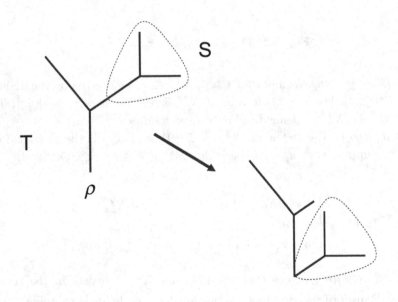

Fig. 5.3. Pruning off the subtree S and regrafting it at the root ρ

Now consider two other finite, rooted subtrees (T^{**}, ρ) and (T^{***}, ρ) of T such that $T^* \cup T^{**} \subseteq T^{***}$ (with induced metrics).

Build $X^{T^{**}}$ and $X^{T^{***}}$ from π_0 and π in the same manner as X^{T^*} (but starting at T^{**} and T^{***}). It is clear from the construction that:

- $X_t^{T^*}$ and $X_t^{T^{**}}$ are rooted subtrees of $X_t^{T^{***}}$ for all $t \geqslant 0$,
- the Hausdorff distance between $X_t^{T^*}$ and $X_t^{T^{**}}$ as subsets of $X_t^{T^{***}}$ does not depend on T^{***},
- the Hausdorff distance is constant between jumps of X^{T^*} and $X^{T^{**}}$ (when only root growth is occurring in both processes).

The following lemma shows that the Hausdorff distance between $X_t^{T^*}$ and $X_t^{T^{**}}$ as subsets of $X_t^{T^{***}}$ does not increase at jump times.

Lemma 5.4. *Let T be a finite rooted tree with root ρ and metric d, and let T' and T'' be two rooted subtrees of T (both with the induced metrics and root ρ). Fix $p \in T$, and let S be the subtree in T above p (recall (5.2)). Define a new metric $\hat{d}$ on T by putting*

$$\hat{d}(a,b) := \begin{cases} d(a,b), & \text{if } a, b \in S, \\ d(a,b), & \text{if } a, b \in T \backslash S, \\ d(a,p) + d(\rho, b), & \text{if } a \in S, b \in T \backslash S. \end{cases}$$

Then the sets T' and T'' are also subtrees of T equipped with the induced metric $\hat{d}$, and the Hausdorff distance between T' and T'' with respect to $\hat{d}$ is not greater than that with respect to d.

Proof. Suppose that the Hausdorff distance between T' and T'' under d is less than some given $\varepsilon > 0$. Given $a \in T'$, there then exists $b \in T''$ such that $d(a,b) < \varepsilon$. Because $d(a, a \wedge b) \leqslant d(a,b)$ and $a \wedge b \in T''$, we may suppose (by replacing b by $a \wedge b$ if necessary) that $b \leqslant a$.

We claim that $\hat{d}(a,c) < \varepsilon$ for some $c \in T''$. This and the analogous result with the roles of T' and T'' interchanged will establish the result.

If $a, b \in S$ or $a, b \in T \backslash S$, then $\hat{d}(a,b) = d(a,b) < \varepsilon$. The only other possibility is that $a \in S$ and $b \in T \backslash S$, in which case $p \in [b, a]$ (for T equipped with d). Then $\hat{d}(a, \rho) = d(a,p) \leqslant d(a,b) < \varepsilon$, as required (because $\rho \in T''$). $\qquad \square$

Now let $T_1 \subseteq T_2 \subseteq \cdots$ be an increasing sequence of finite subtrees of T such that $\bigcup_{n \in \mathbb{N}} T_n$ is dense in T. Thus, $\lim_{n \to \infty} d_H(T_n, T) = 0$.

Let $X^1, X^2, \ldots$ be constructed from π_0 and π starting with $T_1, T_2, \ldots$. Applying Lemma 5.4 yields

$$\lim_{m,n \to \infty} \sup_{t \geqslant 0} d_{\mathrm{GH^{root}}}(X_t^m, X_t^n) = 0.$$

Hence, by completeness of $\mathbf{T}^{\mathrm{root}}$, there exists a càdlàg $\mathbf{T}^{\mathrm{root}}$-valued process X such that $X_0 = T$ and

$$\limsup_{m \to \infty} d_{\mathrm{GH^{root}}}(X_t^m, X_t) = 0.$$

A priori, the process X could depend on the choice of the approximating sequence of trees $\{T_n\}_{n \in \mathbb{N}}$. To see that this is not so, consider two approximating sequences $T_1^1 \subseteq T_2^1 \subseteq \cdots$ and $T_1^2 \subseteq T_2^2 \subseteq \cdots$.

For $k \in \mathbb{N}$, write T_n^3 for the smallest rooted subtree of T that contains both T_n^1 and T_n^2. As a set, $T_n^3 = T_n^1 \cup T_n^2$. Now let $\{(X_t^{n,i}\}_{t \geqslant 0})_{n \in \mathbb{N}}$ for $i = 1, 2, 3$ be the corresponding sequences of finite tree-value processes and let $(X_t^{\infty,i})_{t \geqslant 0}$ for $i = 1, 2, 3$ be the corresponding limit processes. By Lemma 5.4,

$$
\begin{aligned}
d_{\mathrm{GH^{root}}}(X_t^{n,1}, X_t^{n,2}) &\leqslant d_{\mathrm{GH^{root}}}(X_t^{n,1}, X_t^{n,3}) + d_{\mathrm{GH^{root}}}(X_t^{n,2}, X_t^{n,3}) \\
&\leqslant d_{\mathrm{H}}(X_t^{n,1}, X_t^{n,3}) + d_{\mathrm{H}}(X_t^{n,2}, X_t^{n,3}) \\
&\leqslant d_{\mathrm{H}}(T_n^1, T_n^3) + d_{\mathrm{H}}(T_n^2, T_n^3) \\
&\leqslant d_{\mathrm{H}}(T_n^1, T) + d_{\mathrm{H}}(T_n^2, T) \to 0
\end{aligned}
\tag{5.3}
$$

as $n \to \infty$.

Thus, for each $t \geqslant 0$ the sequences $\{X_t^{n,1}\}_{n \in \mathbb{N}}$ and $\{X_t^{n,2}\}_{n \in \mathbb{N}}$ do indeed have the same rooted Gromov–Hausdorff limit and the process X does not depend on the choice of approximating sequence for the initial tree T.

5.2.3 Putting Randomness into the Construction

We constructed a $\mathbf{T}^{\mathrm{root}}$-valued function $t \mapsto X_t$ starting with a fixed triple (T, π_0, π), where $T \in \mathbf{T}^{\mathrm{root}}$ and π_0, π satisfy the conditions of Assumption 5.2. We now want to think of X as a function of time and such triples.

Let Ω^* be the set of triples (T, π_0, π), where T is a rooted compact $\mathbb{R}$-tree (that is, a class representative of an element of $\mathbf{T}^{\mathrm{root}}$) and π_0, π satisfy Assumption 5.2.

The root invariant isometry equivalence relation on rooted compact $\mathbb{R}$-trees extends naturally to an equivalence relation on Ω^* by declaring that two triples (T', π_0', π') and (T'', π_0'', π''), where $\pi_0' = \{(\sigma_i', x_i') : i \in \mathbb{N}\}$ and $\pi_0'' = \{(\sigma_i'', x_i'') : i \in \mathbb{N}\}$, are equivalent if there is a root invariant isometry f mapping T' to T'' and a permutation γ of $\mathbb{N}$ such that $\sigma_i'' = \sigma_{\gamma(i)}'$ and $x_i'' = f(x_{\gamma(i)}')$ for all $i \in \mathbb{N}$. Write Ω for the resulting quotient space of equivalence classes. There is a natural measurable structure on Ω: we refer to [63] for the details.

Given $T \in \mathbf{T}^{\mathrm{root}}$, let $\mathbf{P}^T$ be the probability measure on Ω defined by the following requirements.

- The measure $\mathbf{P}^T$ assigns all of its mass to the set $\{(T', \pi_0', \pi') \in \Omega : T' = T\}$.
- Under $\mathbf{P}^T$, the random variable $(T', \pi_0', \pi') \mapsto \pi_0'$ is a Poisson point process on the set $\mathbb{R}^{++} \times T^o$ with intensity $\lambda \otimes \mu$, where μ is the length measure on T.

- Under $\mathbf{P}^T$, the random variable $(T', \pi'_0, \pi') \mapsto \pi'$ is a Poisson point process on the set $\{(t, x) \in \mathbb{R}^{++} \times \mathbb{R}^{++} : x \leqslant t\}$ with intensity $\lambda \otimes \lambda$ restricted to this set.
- The random variables $(T', \pi'_0, \pi') \mapsto \pi'_0$ and $(T', \pi'_0, \pi') \mapsto \pi'$ are independent under $\mathbf{P}^T$.

Of course, the random variable $(T', \pi'_0, \pi') \mapsto \pi'_0$ takes values in a space of equivalence classes of countable sets rather than a space of sets *per se*, so, more formally, this random variable has the law of the image of a Poisson process on an arbitrary class representative under the appropriate quotient map.

For $t \geqslant 0$, g a bounded Borel function on $\mathbf{T}^{\mathrm{root}}$, and $T \in \mathbf{T}^{\mathrm{root}}$, set

$$P_t g(T) := \mathbf{P}^T[g(X_t)]. \tag{5.4}$$

With a slight abuse of notation, let $\tilde{R}_\eta$ for $\eta > 0$ also denote the map from Ω into Ω that sends (T, π_0, π) to $(R_\eta(T), \pi_0 \cap (\mathbb{R}^{++} \times (R_\eta(T))^\circ), \pi)$.

Theorem 5.5. *(i) If $T \in \mathbf{T}^{\mathrm{root}}$ is finite, then $(X_t)_{t \geqslant 0}$ under $\mathbf{P}^T$ is a Markov process that evolves via the root growth with re-grafting dynamics on finite trees.*

(ii) For all $\eta > 0$ and $T \in \mathbf{T}^{\mathrm{root}}$, the law of $(X_t \circ \tilde{R}_\eta)_{t \geqslant 0}$ under $\mathbf{P}^T$ coincides with the law of $(X_t)_{t \geqslant 0}$ under $\mathbf{P}^{R_\eta(T)}$.

(iii) For all $T \in \mathbf{T}^{\mathrm{root}}$, the law of $(X_t)_{t \geqslant 0}$ under $\mathbf{P}^{R_\eta(T)}$ converges as $\eta \downarrow 0$ to that of $(X_t)_{t \geqslant 0}$ under $\mathbf{P}^T$ (in the sense of convergence of laws on the space of càdlàg $\mathbf{T}^{\mathrm{root}}$-valued paths equipped with the Skorohod topology).

(iv) For $g \in \mathrm{b}\mathcal{B}(\mathbf{T}^{\mathrm{root}})$, the map $(t, T) \mapsto P_t g(T)$ is $\mathcal{B}(\mathbb{R}^+) \times \mathcal{B}(\mathbf{T}^{\mathrm{root}})$-measurable.

(v) The process $(X_t, \mathbf{P}^T)$ is strong Markov and has transition semigroup $(P_t)_{t \geqslant 0}$.

Proof. (i) This is clear from the definition of the root growth and re-grafting dynamics.

(ii) It is enough to check that the push-forward of the probability measure $\mathbf{P}^T$ under the map $R_\eta : \Omega \to \Omega$ is the measure $\mathbf{P}^{R_\eta(T)}$.

This, however, follows from the observation that the restriction of length measure on a tree to a subtree is just length measure on the subtree.

(iii) This is immediate from part (ii) and part (iv) of Lemma 4.32. Indeed, we have that

$$\sup_{t \geqslant 0} d_{\mathrm{GH}^{\mathrm{root}}}(X_t, X_t \circ \tilde{R}_\eta) \leqslant d_{\mathrm{H}}(T, R_\eta(T)) \leqslant \eta.$$

(iv) By a monotone class argument, it is enough to consider the case where the test function g is continuous. It follows from part (iii) that $P_t g(R_\eta(T))$ converges pointwise to $P_t g(T)$ as $\eta \downarrow 0$, and it is not difficult to show using Lemma 4.32 and part (i) that $(t, T) \mapsto P_t g(R_\eta(T))$ is $\mathcal{B}(\mathbb{R}^+) \times \mathcal{B}(\mathbf{T}^{\mathrm{root}})$-measurable, but we omit the details.

(v) By construction and Lemma 4.33, we have for $t \geqslant 0$ and $(T, \pi_0, \pi) \in \Omega$ that, as a set, $X_t^o(T, \pi_0, \pi)$ is the disjoint union $T^o \sqcup]0, t]$.

Put

$$
\theta_t(T, \pi_0, \pi)
$$
$$
:= \Big(X_t(T, \pi_0, \pi), \{(s, x) \in \mathbb{R}^{++} \times T^o : (t + s, x) \in \pi_0\},
$$
$$
\{(s, x) \in \mathbb{R}^{++} \times \mathbb{R}^{++} : (t + s, t + x) \in \pi\} \Big)
$$
$$
= \Big(X_t(T, \pi_0, \pi), \{(s, x) \in \mathbb{R}^{++} \times X_t^o(T, \pi_0, \pi) : (t + s, x) \in \pi_0\},
$$
$$
\{(s, x) \in \mathbb{R}^{++} \times \mathbb{R}^{++} : (t + s, t + x) \in \pi\} \Big).
$$

Thus, θ_t maps Ω into Ω. Note that $X_s \circ \theta_t = X_{s+t}$ and that $\theta_s \circ \theta_t = \theta_{s+t}$, that is, the family $(\theta_t)_{t \geqslant 0}$ is a semigroup.

Fix $t \geqslant 0$ and $(T, \pi_0, \pi) \in \Omega$. Write μ' for the measure on $T^o \sqcup]0, t]$ that restricts to length measure on T^o and to Lebesgue measure on $]0, t]$. Write μ'' for the length measure on $X_t^o(T, \pi_0, \pi)$.

The strong Markov property will follow from a standard strong Markov property for Poisson processes if we can show that $\mu' = \mu''$.

This equality is clear from the construction if T is finite: the tree $X_t(T, \pi_0, \pi)$ is produced from the tree T and the set $]0, t]$ by a finite number of dissections and rearrangements.

The equality for general T follows from the construction and Lemma 4.33.

$\square$

5.2.4 Feller Property

The proof of Theorem 5.5 depended on an argument that showed that if we have two finite subtrees of a given tree that are close in the Gromov–Hausdorff distance, then the resulting root growth with re-grafting processes can be coupled together on the same probability space so that they stay close together. It is believable that if we start the root growth with re-grafting process with any two trees that are close together (whether or not they are finite or subtrees of of a common tree), then the resulting processes will be close in some sense. The following result, which implies that the measure induced by the root growth with re-grafting process on path space is weakly continuous in the starting state with respect to the Skorohod topology on path space can be established by a considerably more intricate coupling argument: we refer to [63] for the details.

Proposition 5.6. *If the function $f : \mathbf{T}^{\text{root}} \to \mathbb{R}$ is continuous and bounded, then the function $P_t f$ is also continuous and bounded for each $t \geqslant 0$.*

5.3 Ergodicity, Recurrence, and Uniqueness

5.3.1 Brownian CRT and Root Growth with Re-Grafting

Recall that Algorithm 2.4 for generating uniform rooted tree on n labeled vertices was derived from Algorithm 5.1, the tree-valued Markov chain appearing in the proof of the Markov chain tree theorem that has the uniform rooted tree on n labeled vertices as its stationary distribution. Recall also that the Poisson line-breaking construction of the Brownian continuum random dom tree in Section 2.5 is an asymptotic version of Algorithm 2.4, whilst the root growth with re-grafting process was motivated as an asymptotic version of Algorithm 5.1. Therefore, it seems reasonable that there should be a connection between the Poisson line-breaking construction and the root growth with re-grafting process. We establish the connection in this subsection.

Let us first present the Poisson line-breaking construction in a more "dynamic" way that will make the comparison with the root growth with re-grafting process a little more transparent.

- Write $\tau_1, \tau_2, \ldots$ for the successive arrival times of an inhomogeneous Poisson process with arrival rate t at time $t \geqslant 0$. Call τ_n the nth *cut time*.
- Start at time 0 with the 1-tree (that is a line segment with two ends), $\mathcal{R}_0$, of length zero ($\mathcal{R}_0$ is "really" the *trivial tree* that consists of one point only, but thinking this way helps visualize the dynamics more clearly for this semi-formal description). Identify one end of $\mathcal{R}_0$ as the root.
- Let this line segment grow at unit speed until the first cut time τ_1.
- At time τ_1 pick a point uniformly on the segment that has been grown so far. Call this point the *first cut point*.
- Between time τ_1 and time τ_2, evolve a tree with 3 ends by letting a new branch growing away from the first cut point at unit speed.
- Proceed inductively: Given the n-tree (that is, a tree with $n+1$ ends), $\mathcal{R}_{\tau_n-}$, pick the *n-th cut point* uniformly on $\mathcal{R}_{\tau_n-}$ to give an $n + 1$-tree, $\mathcal{R}_{\tau_n}$, with one edge of length zero, and for $t \in [\tau_n, \tau_{n+1}[$, let $\mathcal{R}_t$ be the tree obtained from $\mathcal{R}_{\tau_n}$ by letting a branch grow away from the nth cut point with unit speed.

The tree $\mathcal{R}_{\tau_n-}$ is n^{th} step of the Poisson line-breaking construction, and the Brownian CRT is the limit of the increasing family of rooted finite trees $(\mathcal{R}_t)_{t \geqslant 0}$.

We will now use the ingredients appearing in the construction of $\mathcal{R}$ to construct a version of the root growth with re-grafting process started at the trivial tree.

- Let $\tau_1, \tau_2, \ldots$ be as in the construction of the $\mathcal{R}$.
- Start with the 1-tree (with one end identified as the *root* and the other as a *leaf*), $\mathcal{T}_0$, of length zero.

- Let this segment grow at unit speed on the time interval $[0, \tau_1[$, and for $t \in [0, \tau_1[$ let $\mathcal{T}_t$ be the rooted 1-tree that has its points labeled by the interval $[0, t]$ in such a way that the root is t and the leaf is 0.
- At time τ_1 sample the *first cut point* uniformly along the tree $\mathcal{T}_{\tau_1-}$, prune off the piece of $\mathcal{T}_{\tau_1-}$ that is above the cut point (that is, prune off the interval of points that are further away from the root t than the first cut point).
- Re-graft the pruned segment such that its cut end and the root are glued together. Just as we thought of $\mathcal{T}_0$ as a tree with two points, (a leaf and a root) connected by an edge of length zero, we take $\mathcal{T}_{\tau_1}$ to be the the rooted 2-tree obtained by "ramifying" the root $\mathcal{T}_{\tau_1-}$ into two points (one of which we keep as the root) that are joined by an edge of length zero.
- Proceed inductively: Given the labeled and rooted n-tree, $\mathcal{T}_{\tau_{n-1}}$, for $t \in [\tau_{n-1}, \tau_n[$, let $\mathcal{T}_t$ be obtained by letting the edge containing the root grow at unit speed so that the points in $\mathcal{T}_t$ correspond to the points in the interval $[0, t]$ with t as the root. At time τ_n, the nth cut point is sampled randomly along the edges of the n-tree, $\mathcal{T}_{\tau_n-}$, and the subtree above the cut point (that is the subtree of points further away from the root than the cut point) is pruned off and re-grafted so that its cut end and the root are glued together. The root is then "ramified" as above to give an edge of length zero leading from the root to the rest of the tree.

Let $(\mathcal{R}_t)_{t \geq 0}$, $(\mathcal{T}_t)_{t \geq 0}$, and $\{\tau_n\}_{n \in \mathbb{N}}$ be as above. Note that $(\mathcal{T}_t)_{t \geq 0}$ has the same law as $(X_t)_{t \geq 0}$ under $\mathbf{P}^{T_0}$, where T_0 is the trivial tree.

Proposition 5.7. *The two random finite rooted trees $\mathcal{R}_{\tau_n-}$ and $\mathcal{T}_{\tau_n-}$ have the same distribution for all $n \in \mathbb{N}$.*

Proof. Let R_n denote the object obtained by taking the rooted finite tree with edge lengths $\mathcal{R}_{\tau_n-}$ and labeling the leaves with $1, \ldots, n$, in the order that they are added in Aldous's construction. Let T_n be derived similarly from the rooted finite tree with edge lengths $\mathcal{T}_{\tau_n-}$, by labeling the leaves with $1, \ldots, n$ in the order that they appear in the root growth with re-grafting construction. It will suffice to show that R_n and T_n have the same distribution. Note that both R_n and T_n are rooted bifurcating trees with n labeled leaves and edge lengths. Such a tree S_n is uniquely specified by its *shape*, denoted shape(S_n), that is a rooted, bifurcating, leaf-labeled combinatorial tree, and by the list of its $(2n - 1)$ edge lengths in a canonical order determined by its shape, say

$$\text{lengths}(S_n) := (\text{length}(S_n, 1), \ldots, \text{length}(S_n, 2n - 1)),$$

where the edge lengths are listed in order of traversal of edges by first working along the path from the root to leaf 1, then along the path joining that path to leaf 2, and so on.

Recall that τ_n is the nth point of a Poisson process on $\mathbb{R}^{++}$ with rate $t\,dt$. We construct R_n and T_n on the same probability space using cuts at points $U_i\tau_i$, $1 \leqslant i \leqslant n - 1$, where $U_1, U_2, \ldots$ is a sequence of independent random variables uniformly distributed on the interval $]0, 1]$ and independent of the sequence $\{\tau_n\}_{n\in\mathbb{N}}$. Then, by construction, the common collection of edge lengths of R_n and of T_n is the collection of lengths of the $2n - 1$ subintervals of $]0, \tau_n]$ obtained by cutting this interval at the $2n - 2$ points

$$\{X_i^{(n)} : 1 \leqslant i \leqslant 2n - 2\} := \bigcup_{i=1}^{n-1} \{U_i\tau_i, \tau_i\}$$

where the $X_i^{(n)}$ are indexed to increase in i for each fixed n. Let $X_0^{(n)} := 0$ and $X_{2n-1}^{(n)} := \tau_n$. Then

$$\text{length}(R_n, i) = X_i^{(n)} - X_{i-1}^{(n)}, \quad 1 \leqslant i \leqslant 2n - 1, \qquad (5.5)$$

$$\text{length}(T_n, i) = \text{length}(R_n, \sigma_{n,i}), \quad 1 \leqslant i \leqslant 2n - 1, \qquad (5.6)$$

for some almost surely unique random indices $\sigma_{n,i} \in \{1, \ldots 2n - 1\}$ such that $i \mapsto \sigma_{n,i}$ is almost surely a permutation of $\{1, \ldots 2n - 1\}$. According to [10, Lemma 21], the distribution of R_n may be characterized as follows:

(i) the sequence lengths(R_n) is exchangeable, with the same distribution as the sequence of lengths of subintervals obtained by cutting $]0, \tau_n]$ at $2n - 2$ uniformly chosen points $\{U_i\tau_n : 1 \leqslant i \leqslant 2n - 2\}$;

(ii) shape(R_n) is uniformly distributed on the set of all $1 \times 3 \times 5 \times \cdots \times (2n-3)$ possible shapes;

(iii) lengths(R_n) and shape(R_n) are independent.

In view of this characterization and (5.6), to show that T_n has the same distribution as R_n it is enough to show that

(a) the random permutation $\{i \mapsto \sigma_{n,i} : 1 \leqslant i \leqslant 2n - 1\}$ is a function of shape(T_n);

(b) shape$(T_n) = \Psi_n(\text{shape}(R_n))$ for some bijective map Ψ_n from the set of all possible shapes to itself.

This is trivial for $n = 1$, so we assume below that $n \geqslant 2$. Before proving (a) and (b), we recall that (ii) above involves a natural bijection

$$(I_1, \ldots, I_{n-1}) \leftrightarrow \text{shape}(R_n) \qquad (5.7)$$

where $I_{n-1} \in \{1, \ldots, 2n - 3\}$ is the unique i such that

$$U_{n-1}\tau_{n-1} \in (X_{i-1}^{(n-1)}, X_i^{(n-1)}).$$

Hence, I_{n-1} is the index in the canonical ordering of edges of R_{n-1} of the edge that is cut in the transformation from R_{n-1} to R_n by attachment of an additional edge, of length $\tau_n - \tau_{n-1}$, connecting the cut-point to leaf n. Thus, (ii) and (iii) above correspond via (5.7) to the facts that $I_1, \ldots, I_{n-1}$

are independent and uniformly distributed over their ranges, and independent of lengths(R_n). These facts can be checked directly from the construction of $\{R_n\}_{n \in \mathbb{N}}$ from $\{\tau_n\}_{n \in \mathbb{N}}$ and $\{U_n\}_{n \in \mathbb{N}}$ using standard facts about uniform order statistics.

Now (a) and (b) follow from (5.7) and another bijection

$$(I_1, \ldots, I_{n-1}) \leftrightarrow \text{shape}(T_n) \tag{5.8}$$

where each possible value i of I_m is identified with edge $\sigma_{m,i}$ in the canonical ordering of edges of T_m. This is the edge of T_m whose length equals length(R_m, i). The bijection (5.8), and the fact that $\sigma_{n,i}$ depends only on shape(T_n), will now be established by induction on $n \geq 2$. For $n = 2$ the claim is obvious. Suppose for some $n \geq 3$ that the correspondence between $(I_1, \ldots, I_{n-2})$ and shape(T_{n-1}) has been established, and that the length of edge $\sigma_{n-1,i}$ in the canonical ordering of edges of T_{n-1} is equals the length of the ith edge in the canonical ordering of edges of R_{n-1}, for some $\sigma_{n-1,i}$ that is a function of i and shape(T_{n-1}). According to the construction of T_n, if $I_{n-1} = i$ then T_n is derived from T_{n-1} by splitting T_{n-1} into two branches at some point along edge $\sigma_{n-1,i}$ in the canonical ordering of the edges of T_{n-1}, and forming a new tree from the two branches and an extra segment of length $\tau_n - \tau_{n-1}$. Clearly, shape(T_n) is determined by shape(T_{n-1}) and I_{n-1}, and in the canonical ordering of the edge lengths of T_n the length of the ith edge equals the length of the edge $\sigma_{n,i}$ of R_n, for some $\sigma_{n,i}$ that is a function of shape(T_{n-1}) and I_{n-1}, and, therefore, a function of shape(T_n). To complete the proof, it is enough by the inductive hypothesis to show that the map

$$(\text{shape}(T_{n-1}), I_{n-1}) \rightarrow \text{shape}(T_n)$$

just described is invertible. But shape(T_{n-1}) and I_{n-1} can be recovered from shape(T_n) by the following sequence of moves:

- delete the edge attached to the root of shape(T_n)
- split the remaining tree into its two branches leading away from the internal node to which the deleted edge was attached;
- re-attach the bottom end of the branch not containing leaf n to leaf n on the other branch, joining the two incident edges to form a single edge;
- the resulting shape is shape(T_{n-1}), and I_{n-1} is the index such that the joined edge in shape(T_{n-1}) is the edge $\sigma_{n-1,I_{n-1}}$ in the canonical ordering of edges on shape(T_{n-1}).

$\square$

5.3.2 Coupling

Lemma 5.8. *For any $(T, d, \rho) \in \mathbf{T}^{\text{root}}$ we can build on the same probability space two $\mathbf{T}^{\text{root}}$-valued processes X' and X'' such that:*

- X' has the law of X under $\mathbf{P}^{T_0}$, where T_0 is the trivial tree consisting of just the root,
- X'' has the law of X under $\mathbf{P}^T$,
- for all $t \geqslant 0$,

$$d_{\mathrm{GH^{root}}}(X'_t, X''_t) \leqslant d_{\mathrm{GH^{root}}}(T_0, T) = \sup\{d(\rho, x) : x \in T\} \qquad (5.9)$$

-

$$\lim_{t \to \infty} d_{\mathrm{GH^{root}}}(X'_t, X''_t) = 0, \quad \textit{almost surely.} \qquad (5.10)$$

Proof. The proof follows almost immediately from construction of X and Lemma 5.4. The only point requiring some comment is (5.10).

For that it will be enough to show for any $\varepsilon > 0$ that for $\mathbf{P}^T$-a.e. $(T, \pi_0, \pi) \in \Omega$ there exists $t > 0$ such that the projection of $\pi_0 \cap (]0, t] \times T^o)$ onto T is an ε-net for T.

Note that the projection of $\pi_0 \cap (]0, t] \times T^o)$ onto T is a Poisson process under $\mathbf{P}^T$ with intensity $t\mu$, where μ is the length measure on T. Moreover, T can be covered by a finite collection of ε-balls, each with positive μ-measure.

Therefore, the $\mathbf{P}^T$-probability of the set of $(T, \pi_0, \pi) \in \Omega$ such that the projection of $\pi_0 \cap (]0, t] \times T^o)$ onto T is an ε-net for T increases as $t \to \infty$ to 1. $\qquad \square$

5.3.3 Convergence to Equilibrium

Proposition 5.9. *For any $T \in \mathbf{T}^{\mathrm{root}}$, the law of X_t under $\mathbf{P}^T$ converges weakly to that of the Brownian CRT as $t \to \infty$.*

Proof. It suffices by Lemma 5.8 to consider the case where T is the trivial tree.

We saw in the Proposition 5.7 that, in the notation of that result, $\mathcal{T}_{\tau_n-}$ has the same distribution as $\mathcal{R}_{\tau_n-}$.

Moreover, $\mathcal{R}_t$ converges in distribution to the continuum random tree as $t \to \infty$ if we use Aldous's metric on trees that comes from thinking of them as closed subsets of ℓ^1 with the root at the origin and equipped with the Hausdorff distance.

By construction, $(\mathcal{T}_t)_{t \geqslant 0}$ has the root growth with re-grafting dynamics started at the trivial tree. Clearly, the rooted Gromov–Hausdorff distance between $\mathcal{T}_t$ and $\mathcal{T}_{\tau_{n+1}-}$ is at most $\tau_{n+1} - \tau_n$ for $\tau_n \leqslant t < \tau_{n+1}$.

It remains to observe that $\tau_{n+1} - \tau_n \to 0$ in probability as $n \to \infty$. $\qquad \square$

5.3.4 Recurrence

Proposition 5.10. *Consider a non-empty open set $U \subseteq \mathbf{T}^{\mathrm{root}}$. For each $T \in \mathbf{T}^{\mathrm{root}}$,*

$$\mathbf{P}^T\{\textit{for all } s \geqslant 0, \textit{ there exists } t > s \textit{ such that } X_t \in U\} = 1. \qquad (5.11)$$

Proof. It is straightforward, but notationally rather tedious, to show that if $B' \subseteq \mathbf{T}^{\mathrm{root}}$ is any ball and T_0 is the trivial tree, then

$$\mathbf{P}^{T_0}\{X_t \in B'\} > 0 \tag{5.12}$$

for all t sufficiently large.

Thus, for any ball $B' \subseteq \mathbf{T}^{\mathrm{root}}$ there is, by Lemma 5.8, a ball $B'' \subseteq \mathbf{T}^{\mathrm{root}}$ containing the trivial tree such that

$$\inf_{T \in B''} \mathbf{P}^T\{X_t \in B'\} > 0 \tag{5.13}$$

for each t sufficiently large.

By a standard application of the Markov property, it therefore suffices to show for each $T \in \mathbf{T}^{\mathrm{root}}$ and each ball B'' around the trivial tree that

$$\mathbf{P}^T\{\text{there exists } t > 0 \text{ such that } X_t \in B''\} = 1. \tag{5.14}$$

By another standard application of the Markov property, equation (5.14) will follow if we can show that there is a constant $p > 0$ depending on B'' such that for any $T \in \mathbf{T}^{\mathrm{root}}$

$$\liminf_{t \to \infty} \mathbf{P}^T\{X_t \in B''\} > p.$$

This, however, follows from Proposition 5.9 and the observation that for any $\varepsilon > 0$ the law of the Brownian CRT assigns positive mass to the set of trees with height less than ε: this is just the observation that the law of the Brownian excursion assigns positive mass to the set of excursion paths with maximum less that $\varepsilon/2$.   □

5.3.5 Uniqueness of the Stationary Distribution

Proposition 5.11. *The law of the Brownian CRT is the unique stationary distribution for X. That is, if ξ is the law of the CRT, then*

$$\int \xi(dT) P_t f(T) = \int \xi(dT) f(T)$$

for all $t \geq 0$ and $f \in b\mathcal{B}(\mathbf{T}^{\mathrm{root}})$, and ξ is the unique probability measure on $\mathbf{T}^{\mathrm{root}}$ with this property.

Proof. This is a standard argument given Proposition 5.9 and the Feller property for the semigroup $(P_t)_{t \geq 0}$ established in Proposition 5.6, but we include the details for completeness.

Consider a test function $f : \mathbf{T}^{\mathrm{root}} \to \mathbb{R}$ that is continuous and bounded. By Proposition 5.6, the function $P_t f$ is also continuous and bounded for each $t \geq 0$.

Therefore, by Proposition 5.9,

$$\int \xi(dT)f(T) = \lim_{s\to\infty} \int \xi(dT)P_s f(T) = \lim_{s\to\infty} \int \xi(dT)P_{s+t}f(T)$$

$$= \lim_{s\to\infty} \int \xi(dT)P_s(P_t f)(T) = \int \xi(dT)P_t f(T)$$

(5.15)

for each $t \geqslant 0$. Hence, ξ is stationary.

Moreover, if ζ is a stationary measure, then

$$\int \zeta(dT)f(T) = \int \zeta(dT)P_t f(T)$$

$$\to \int \zeta(dT)\left(\int \xi(dT)f(T)\right) = \int \xi(dT)f(T),$$

(5.16)

and $\zeta = \xi$, as claimed. $\qquad\square$

5.4 Convergence of the Markov Chain Tree Algorithm

We would like to show that Algorithm 5.1 converges to a process having the root growth with re-grafting dynamics after suitable re-scaling of time and edge lengths of the evolving tree. It will be more convenient for us to work with the continuous time version of the algorithm in which the transitions are made at the arrival times of an independent Poisson process with rate 1.

The continuous time version of Algorithm 5.1 involves a labeled combinatorial tree, but, by symmetry, if we don't record the labeling and associate rooted labeled combinatorial trees with rooted compact real trees having edges that are line segments with length 1, then the resulting process will still be Markovian.

It will be convenient to use the following notation for re-scaling the distances in a $\mathbb{R}$-tree: $T = (T, d, \rho)$ is a rooted compact real tree and $c > 0$, we write cT for the tree (T, cd, ρ) (that is, $cT = T$ as sets and the roots are the same, but the metric is re-scaled by c).

Proposition 5.12. *Let $Y^n = (Y_t^n)_{t\geqslant 0}$ be a sequence of Markov processes that take values in the space of rooted compact real trees with integer edge lengths and evolve according to the dynamics associated with the continuous-time version of Algorithm 5.1. Suppose that each tree Y_0^n is non-random with total branch length N_n, that N_n converges to infinity as $n \to \infty$, and that $N_n^{-1/2}Y_0^n$ converges in the rooted Gromov–Hausdorff metric to some rooted compact real tree T as $n \to \infty$. Then, in the sense of weak convergence of processes on the space of càdlàg paths equipped with the Skorohod topology, $(N_n^{-1/2}Y^n(N_n^{1/2}t))_{t\geqslant 0}$ converges as $n \to \infty$ to the root growth with re-grafting process X under $\mathbf{P}^T$.*

Proof. Define $Z^n = (Z_t^n)_{t\geqslant 0}$ by

$$Z_t^n := N_n^{-1/2}Y^n(N_n^{1/2}t).$$

For $\eta > 0$, let $Z^{\eta,n}$ be the $\mathbf{T}^{\text{root}}$-valued process constructed as follows.

- Set $Z_0^{\eta,n} = R_{\eta_n}(Z_0^n)$, where $\eta_n := N_n^{-1/2}\lfloor N_n^{1/2}\eta \rfloor$.
- The value of $Z^{\eta,n}$ is unchanged between jump times of $(Z_t^n)_{t\geqslant 0}$.
- At a jump time τ for $(Z_t^n)_{t\geqslant 0}$, the tree $Z_\tau^{\eta,n}$ is the subtree of Z_τ^n spanned by $Z_{\tau-}^{\eta,n}$ and the root of Z_τ^n.

An argument similar to that in the proof of Lemma 5.4 shows that

$$\sup_{t\geqslant 0} d_{\mathrm{H}}(Z_t^n, Z_t^{\eta,n}) \leqslant \eta_n,$$

and so it suffices to show that $Z^{\eta,n}$ converges weakly as $n \to \infty$ to X under $\mathbf{P}^{R_\eta(T)}$.

Note that $Z_0^{\eta,n}$ converges to $R_\eta(T)$ as $n \to \infty$. Moreover, if Λ is the map that sends a tree to its total length (that is, the total mass of its length measure), then $\lim_{n\to\infty} \Lambda(Z_0^{\eta,n}) = \Lambda \circ R_\eta(T) < \infty$ by Lemma 4.36 below.

The pure jump process $Z^{\eta,n}$ is clearly Markovian. If it is in a state (T', ρ'), then it jumps with the following rates.

- With rate $N_n^{1/2}(N_n^{1/2}\Lambda(T'))/N_n = \Lambda(T')$, one of the $N_n^{1/2}\Lambda(T')$ points in T' that are at distance a positive integer multiple of $N_n^{-1/2}$ from the root ρ' is chosen uniformly at random and the subtree above this point is joined to ρ' by an edge of length $N_n^{-1/2}$. The chosen point becomes the new root and a segment of length $N_n^{-1/2}$ that previously led from the new root toward ρ' is erased. Such a transition results in a tree with the same total length as T'.
- With rate $N_n^{1/2} - \Lambda(T')$, a new root not present in T' is attached to ρ' by an edge of length $N_n^{-1/2}$. This results in a tree with total length $\Lambda(T') + N_n^{-1/2}$.

It is clear that these dynamics converge to those of the root growth with re-grafting process, with the first class of transitions leading to re-graftings in the limit and the second class leading to root growth.   □

6

The Wild Chain and other Bipartite Chains

6.1 Background

The *wild chain* was introduced informally in Chapter 1. We will now describe it more precisely.

The state space of the wild chain is the set $\mathbf{T}^*$ consisting of rooted $\mathbb{R}$-trees such that each edge has length 1, each vertex has finite degree, and if the tree is infinite there is a single infinite length path from the root. Let μ denote the PGW(1) measure (that is, the distribution of the Galton–Watson tree with mean 1 Poisson offspring distribution) on the set $\mathbf{T}_{<\infty}$ of finite trees in $\mathbf{T}^*$, and let ν denote the distribution of a PGW(1) tree "conditioned to be infinite". It is well-known that ν is concentrated on the set $\mathbf{T}^*_\infty := \mathbf{T}^* \backslash \mathbf{T}_{<\infty}$ consisting of infinite trees with a single infinite path from the root. A realization of ν may be constructed by taking a semi-infinite path, thought of as infinitely many vertices connected by edges of length 1 and appending independent realizations of μ at each vertex. When started in a finite tree from $\mathbf{T}_{<\infty}$, at rate one for each vertex the wild chain attaches that vertex by an edge to the root of a realization of ν. Conversely (and somewhat heuristically), when started in an infinite tree from $\mathbf{T}^*_\infty$, at rate one for each vertex the wild chain prunes off and discards the infinite subtree above that vertex, leaving a finite tree.

The set of times when the state of the wild chain is an infinite tree has Lebesgue measure zero, but it is the uncountable set of points of increase of a continuous additive functional (so that it looks qualitatively like the zero set of a Brownian motion).

The aim of this chapter is to use Dirichlet form methods to construct and study a general class of symmetric Markov processes on a generic totally disconnected state space. Specializing this construction leads to a class of processes that we call *bipartite chains*. This class contains the wild chain as a special case.

In general, we take the state space of the processes we construct to be a Lusin space E such that there exists a countable algebra $\mathcal{R}$ of simultaneously closed and open subsets of E that is a base for the topology of E. Note that E is indeed totally disconnected – see Theorem 33.B of [129]. Conversely, if E is any totally disconnected compact metric space, then there exists a collection $\mathcal{R}$ with the required properties – see Theorem 2.94 of [85].

The following are two instances of such spaces. More examples, including an arbitrary local field and the compactification of an infinite tree, are described in Section 6.2.

Example 6.1. Let E be $\bar{\mathbb{N}} := \mathbb{N} \cup \{\infty\}$, the usual one–point compactification of the positive integers $\mathbb{N} := \{1, 2, \ldots\}$. Equip E with the usual total order and let $\mathcal{R}$ be the algebra generated by sets of the form $\{y : x \leqslant y\}$, $x \in \mathbb{N}$. That is, $\mathcal{R}$ consists of finite subsets of $\mathbb{N}$ and sets that contain a subset of the form $\{z, z + 1, z + 2, \ldots, \infty\}$ for $z \in \mathbb{N}$.

Example 6.2. Let E be the collection $\mathbf{T}_{\leqslant \infty}$ of rooted trees with every vertex having finite degree. Write $\mathbf{T}_{\leqslant n}$ for the subset of $\mathbf{T}_{\leqslant \infty}$ consisting of trees with height at most n. For $m > n$, there is a natural projection map from $\rho_{mn} : \mathbf{T}_{\leqslant m} \to \mathbf{T}_{\leqslant n}$ that throws away vertices of height greater than n and the edges leading to them. We can identify $\mathbf{T}_{\leqslant \infty}$ with the projective limit of this projective system and give it the corresponding projective limit topology (each $\mathbf{T}_{\leqslant n}$ is given the discrete topology), so that $\mathbf{T}_{\leqslant \infty}$ is Polish. Equip $\mathbf{T}_{\leqslant \infty}$ with the inclusion partial order (that is, $x \leqslant y$ if x is a sub–tree of y). Let $\mathcal{R}$ be the algebra generated by sets of the form $\{y : x \leqslant y\}$, $x \in \mathbf{T}_{<\infty} := \bigcup_n \mathbf{T}_{\leqslant n}$. Equivalently, if $\rho_n : \mathbf{T}_{\leqslant \infty} \to \mathbf{T}_{\leqslant n}$ is the projection map that throws away vertices of height greater than n and the edges leading to them, then $\mathcal{R}$ is the collection of sets of the form $\rho_n^{-1}(B)$ for finite or co-finite $B \subseteq \mathbf{T}_{\leqslant n}$, as n ranges over $\mathbb{N}$.

Our main existence result is the following. We prove it in Section 6.3. Appendix A contains a summary of the relevant Dirichlet form theory.

Notation 6.3. *Denote by $\mathcal{C}$ the subalgebra of $bC(E)$ (:= continuous bounded functions on E) generated by the indicator functions of sets in $\mathcal{R}$.*

Theorem 6.4. *Consider two probability measures μ and ν on E and a non-negative Borel function κ on $E \times E$. Define a σ-finite measure Λ on $E \times E$ by $\Lambda(dx, dy) := \kappa(x, y)\mu(dx)\nu(dy)$. Suppose that the following hold:*

(a) the closed support of the measure μ is E;
(b) $\Lambda([(E \backslash R) \times R] \cup [R \times (E \backslash R)]) < \infty$ for all $R \in \mathcal{R}$;
(c) $\int \kappa(x, y) \, \mu(dx) = \infty$ for ν_s-a.e. y, where ν_s is the singular component in the Lebesgue decomposition of ν with respect to μ;
(d) there exists a sequence $(R_n)_{n \in \mathbb{N}}$ of sets in $\mathcal{R}$ such that $\bigcap_{m=n}^{\infty} R_m$ is compact for all n, $\sum_{n \in \mathbb{N}} \mu(E \backslash R_n) < \infty$, and

$$\sum_{n \in \mathbb{N}} \Lambda([(E \backslash R_n) \times R_n] \cup [R_n \times (E \backslash R_n)]) < \infty.$$

Then there is a recurrent μ-symmetric Hunt process $\mathbf{X} = (X_t, \mathbb{P}^x)$ on E whose Dirichlet form is the closure of the form $\mathcal{E}$ on $\mathcal{C}$ defined by

$$\mathcal{E}(f, g) = \iint (f(y) - f(x))(g(y) - g(x))\, \Lambda(dx, dy),\ f, g \in \mathcal{C}.$$

Our standing assumption throughout this chapter is that the conditions of Theorem 6.4 hold.

In order to produce processes that are reminiscent of the wild chain, we need to assume a little more structure on E. Say that E is *bipartite* if there is a countable, dense subset $E^o \subseteq E$ such that each point of E^o is isolated. In particular, E^o is open. In Example 6.1 we can take $E^o = \mathbb{N}$. In Example 6.2 we can take $E^o = T_{<\infty}$. We will see more examples in Section 6.2. Put $E^* = E \backslash E^o$. Note that E^* is the boundary of the open set E^o.

Definition 6.5. *We will call the process $\mathbf{X}$ described in Theorem 6.4 a bipartite Markov chain if the space E is bipartite and, in the notation of Theorem 6.4:*

- *μ is concentrated on E^o,*
- *ν is concentrated on E^*.*

Remark 6.6. For bipartite chains, the measures μ and ν are mutually singular and $\nu_s = \nu$ in the notation of Theorem 6.4. The reference measure μ is invariant for $\mathbf{X}$, that is, $\mathbb{P}^\mu\{X_t \in \cdot\} = \mu$ for each $t \geqslant 0$. Thus, for any $x \in E^o$ we have $\mathbb{P}^x\{X_t \in E^o\} = 1$ for each $t \geqslant 0$, and so $\mathbf{X}$ is Markov chain on the countable set E^o in the same sense that the Feller–McKean chain is a Markov chain on the rationals (the Feller–McKean chain is one-dimensional Brownian motion time-changed by a continuous additive functional that has as its Revuz measure a purely atomic probability measure that assigns positive mass to each rational).

We establish in Proposition 6.14 that the sample–paths of $\mathbf{X}$ bounce backwards and forwards between E^o and E^* in the same manner that the sample paths of the wild chain bounce backwards and forwards between the finite and infinite trees. Also, we show in Proposition 6.16 that under suitable conditions μ is the unique invariant distribution for $\mathbf{X}$ that assigns all of its mass to E^o, and, moreover, for any probability measure γ concentrated on E^o the law of X_t under $\mathbb{P}^\gamma$ converges in total variation to μ as $t \to \infty$.

In Section 6.6 we prove that, in the general setting of Theorem 6.4, the measure ν is the Revuz measure of a positive continuous additive functional (PCAF). We can, therefore, time–change $\mathbf{X}$ using the inverse of this PCAF. When this procedure is applied to a bipartite chain, it produces a Markov process with state space that is a subset of E^*. In particular, we observe in Example 6.24 that instances of this time–change construction lead to "spherically symmetric" Lévy processes on local fields.

A useful tool for proving the last fact is a result from Section 6.5. There we consider a certain type of equivalence relation on E with associated map

π onto the corresponding quotient space. We give conditions on the Dirichlet form $(\mathcal{E}, \mathcal{D}(\mathcal{E}))$ that are sufficient for the process $\pi \circ \mathbf{X}$ to be a symmetric Hunt process.

Notation 6.7. *Write* $(\cdot, \cdot)_\mu$ *for the* $L^2(E, \mu)$ *inner product and* $(T_t)_{t \geqslant 0}$ *for the semigroup on* $L^2(E, \mu)$ *associated with the form* $(\mathcal{E}, \mathcal{D}(\mathcal{E}))$.

6.2 More Examples of State Spaces

Example 6.8. Let E be the usual path–space of a discrete–time Markov chain with countable state–space S augmented by a distinguished cemetery state ∂ to form $S^\partial = S \cup \{\partial\}$. That is, E is the subset of the space of sequences $(S^\partial)^{\mathbb{N}_0}$ (where $\mathbb{N}_0 := \{0, 1, 2, \ldots\}$) consisting of sequences $\{x_n\}_{n \in \mathbb{N}_0}$ such that if $x_n = \partial$ for some n, then $x_m = \partial$ for all $m > n$. Give E the subspace topology inherited from the product topology on $(S^\partial)^{\mathbb{N}_0}$ (where each factor has the discrete topology), so that E is Polish. Given $x \in E$, write $\zeta(x) := \inf\{n : x_n = \partial\} \in \mathbb{N}_0 \cup \{\infty\}$ for the death–time of x. Define a partial order on E by declaring that $x \leqslant y$ if $\zeta(x) \leqslant \zeta(y)$ and $x_n = y_n$ for $0 \leqslant n < \zeta(x)$. (In particular, if x and y are such that $\zeta(x) = \zeta(y) = \infty$, then $x \leqslant y$ if and only if $x = y$.) Let $\mathcal{R}$ be the algebra generated by sets of the form $\{y : x \leqslant y\}$, $\zeta(x) < \infty$. When $\#S = k < \infty$, we can think of E as the regular k-ary rooted tree along with its set of ends. In particular, when $k = 1$ we recover Example 6.1. This example is bipartite with $E^o = \{x : \zeta(x) < \infty\}$,

Example 6.9. A *local field* $\mathbb{K}$ is a locally compact, non-discrete, totally disconnected, topological field. We refer the reader to [135] or [123] for a full discusion of these objects and for proofs of the facts outlined below. More extensive summaries and references to the literature on probability in a local field setting can be found in [58] and [62].

There is a real-valued mapping on $\mathbb{K}$ that we denote by $x \mapsto |x|$. This map, called the *valuation* takes the values $\{q^k : k \in \mathbb{Z}\} \cup \{0\}$, where $q = p^c$ for some prime p and positive integer c and has the properties

$$|x| = 0 \Leftrightarrow x = 0$$
$$|xy| = |x||y|$$
$$|x + y| \leqslant |x| \vee |y|.$$

The mapping $(x, y) \mapsto |x - y|$ on $\mathbb{K} \times \mathbb{K}$ is a metric on $\mathbb{K}$ that gives the topology of K.

Put $\mathbb{D} = \{x : |x| \leqslant 1\}$. The set $\mathbb{D}$ is a ring (the so-called *ring of integers* of K). If we choose $\rho \in K$ so that $|\rho| = q^{-1}$, then

$$\rho^k \mathbb{D} = \{x : |x| \leqslant q^{-k}\} = \{x : |x| < q^{-(k-1)}\}.$$

Every ball is of the form $x + \rho^k \mathbb{D}$ for some $x \in \mathbb{K}$ and $k \in \mathbb{Z}$, and, in particular, all balls are both closed and open. For $\ell < k$ the additive quotient group

$\rho^{\ell}\mathbb{D}/\rho^k\mathbb{D}$ has order $q^{k-\ell}$. Consequently, $\mathbb{D}$ is the union of q disjoint translates of $\rho\mathbb{D}$. Each of these components is, in turn, the union of q disjoint translates of $\rho^2\mathbb{D}$, and so on. Thus, we can think of the collection of balls contained in $\mathbb{D}$ as being arranged in an infinite rooted q-ary tree: the root is $\mathbb{D}$ itself, the nodes at level k are the balls of radius q^{-k} (= cosets of $\rho^k\mathbb{D}$), and the q "children" of such a ball are the q cosets of $\rho^{k+1}\mathbb{D}$ that it contains. We can uniquely associate each point in $\mathbb{D}$ with the sequence of balls that contain it, and so we can think of the points in $\mathbb{D}$ as the ends this tree – see Figure 6.1.

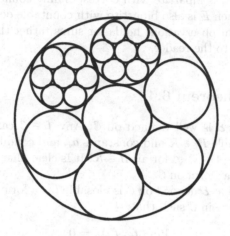

Fig. 6.1. Schematic drawing of the ring of integers $\mathbb{D}$ when $q = p = 7$

This tree picture alone does not capture all the algebraic structure of $\mathbb{D}$; the rings of integers for the p-adic numbers and the p-series field (that is, the field of formal Laurent series with coefficients drawn from the finite field with p elements) are both represented by a p-ary tree, even though the p-adic field has characteristic 0 whereas the p-series field has characteristic p. (As an aside, a locally compact, non-discrete, topological field that is not totally disconnected is necessarily either the real or the complex numbers. Every local field is either a finite algebraic extension of the p-adic number field for some prime p or a finite algebraic extension of the p-series field.)

We can take either $E = \mathbb{K}$ or $E = \mathbb{D}$, with $\mathcal{R}$ the algebra generated by the balls. The same comment applies to Banach spaces over local fields defined as in [123], and we leave the details to the reader.

Example 6.10. In the notation of Example 6.2, let $\mathbf{T}^*_\infty$ be the subset of $\mathbf{T}_{\leqslant\infty}$ consisting of infinite trees through which there is a unique infinite path starting at the root, that is, trees with only one end. Put $\mathbf{T}^* = \mathbf{T}_{<\infty} \cup \mathbf{T}^*_\infty$. It is not hard to see that $E = \mathbf{T}^*$ satisfies our hypothesis, with $\mathcal{R}$ the trace on $\mathbf{T}^*$ of the algebra of subsets of $\mathbf{T}_{\leqslant\infty}$ described in Example 6.2.

Example 6.11. Suppose that the pairs $(E_1, \mathcal{R}_1), \ldots, (E_N, \mathcal{R}_N)$ each satisfy our hypotheses. Put $E := \prod_i E_i$, equip E with the product topology, and set $\mathcal{R}$ to be the algebra generated by subsets of E of the form $\prod_i R_i$ with $R_i \in \mathcal{R}_i$. If each of the factors E_i is bipartite with corresponding countable dense sets of isolated point E_i^o, then E is also bipartite with countable dense set of isolated points $\prod_i E_i^o$. Similar observations holds for sums rather than products, and we leave the details to the reader.

6.3 Proof of Theorem 6.4

We first check that $\mathcal{E}$ is well–defined on $\mathcal{C}$. Any $f \in \mathcal{C}$ can be written $f = \sum_{i=1}^N a_i 1_{R_i}$ for suitable $R_i \in \mathcal{R}$ and constants a_i, and condition (b) is just the condition that $\mathcal{E}(1_R, 1_R) < \infty$ for all $R \in \mathcal{R}$. It is clear that $\mathcal{E}$ is a symmetric, non-negative, bilinear form on $\mathcal{C}$.

We next check that $\mathcal{E}$ defined on $\mathcal{C}$ is closable (as a form on $L^2(E, \mu)$). Let $(f_n)_{n\in\mathbb{N}}$ be a sequence in $\mathcal{C}$ such that

$$\lim_{n\to\infty} (f_n, f_n)_\mu = 0 \qquad (6.1)$$

and

$$\lim_{m,n\to\infty} \mathcal{E}(f_m - f_n, f_m - f_n) = 0. \qquad (6.2)$$

We need to show that

$$\lim_{n\to\infty} \mathcal{E}(f_n, f_n) = 0. \qquad (6.3)$$

Put $\Lambda_s(dx, dy) = \kappa(x, y)\,\mu(dx)\,\nu_s(dy)$. For $M > 0$ put $\Lambda^M(dx, dy) = [\kappa(x, y) \wedge M]\,\mu(dx)\,\nu(dy)$ and $\Lambda_s^M(dx, dy) = [\kappa(x, y) \wedge M]\,\mu(dx)\,\nu_s(dy)$. From (6.1) we have

$$\lim_{m,n\to\infty} \iint (f_m(x) - f_n(x))^2\, \Lambda_s^M(dx, dy) = 0, \ \forall M > 0,$$

and from (6.2) we have

$$\lim_{m,n\to\infty} \iint (\{f_m(y) - f_n(y)\} - \{f_m(x) - f_n(x)\})^2\, \Lambda^M(dx, dy) = 0, \ \forall M > 0.$$
$$(6.4)$$

So, by Minkowski's inequality,

$$\lim_{m,n\to\infty} \iint (f_m(y) - f_n(y))^2 \, \Lambda_s^M(dx, dy) = 0, \quad \forall M > 0. \tag{6.5}$$

Thus, by (6.1), (6.5) and (c), there exists a Borel function f and a sequence $(n_k)_{k\in\mathbb{N}}$ such that $\lim_{k\to\infty} f_{n_k} = 0$, μ-a.e. (and, therefore, ν_a-a.e., where $\nu_a = \nu - \nu_s$ is the absolutely continuous component in the Lebesgue decomposition of ν with respect to μ), and $\lim_{k\to\infty} f_{n_k} = f$, ν_s-a.e.

Now, by Fatou, (6.2) and Minkowski's inequality,

$$\iint f^2(y) \, \Lambda_s(dx, dy) = \iint \lim_{k\to\infty} (f_{n_k}(y) - f_{n_k}(x))^2 \, \Lambda_s(dx, dy)$$

$$\leqslant \liminf_{k\to\infty} \iint (f_{n_k}(y) - f_{n_k}(x))^2 \, \Lambda_s(dx, dy)$$

$$< \infty,$$

and so, by (c), $f = 0$, ν_s-a.e. Finally, by Fatou and (6.2),

$$\lim_{m\to\infty} \iint (f_m(y) - f_m(x))^2 \, \Lambda(dx, dy)$$

$$= \lim_{m\to\infty} \iint \lim_{k\to\infty} (\{f_m(y) - f_{n_k}(y)\} - \{f_m(x) - f_{n_k}(x)\})^2 \, \Lambda(dx, dy)$$

$$\leqslant \lim_{m\to\infty} \liminf_{k\to\infty} \iint (\{f_m(y) - f_{n_k}(y)\} - \{f_m(x) - f_{n_k}(x)\})^2 \, \Lambda(dx, dy)$$

$$= 0,$$

as required.

Write $(\mathcal{E}, \mathcal{D}(\mathcal{E}))$ for the closure of the form $(\mathcal{E}, \mathcal{C})$. To complete the proof that $(\mathcal{E}, \mathcal{D}(\mathcal{E}))$ is a Dirichlet form, it only remains to show that this form is Markov. By Theorem A.7, this will be accomplished if we can show that the unit contraction acts on $(\mathcal{E}, \mathcal{D}(\mathcal{E}))$. That is, we have to show for any $f \in \mathcal{C}$ that

$$(f \vee 0) \wedge 1 \in \mathcal{C} \tag{6.6}$$

and

$$\mathcal{E}((f \vee 0) \wedge 1, (f \vee 0) \wedge 1) \leqslant \mathcal{E}(f, f). \tag{6.7}$$

Considering claim (6.6), first observe that $f \in \mathcal{C}$ if and only if there exist pairwise disjoint $R_1, \ldots, R_N$ and constants $a_1, \ldots, a_N$ such that $f = \sum_i a_i \mathbf{1}_{R_i}$. Thus,

$$(f \wedge 0) \vee 1 = \sum_i ((a_i \vee 0) \wedge 1) \mathbf{1}_{R_i} \in \mathcal{C}.$$

The claim (6.7) is immediate from the definition of $\mathcal{E}$ on $\mathcal{C}$.

We will appeal to Theorem A.8 to establish that $(\mathcal{E}, \mathcal{D}(\mathcal{E}))$ is the Dirichlet form of a μ-symmetric Hunt process, $\mathbf{X}$. It is immediate that conditions (a)–(c) of that result hold for $\mathcal{C}$, so it remains to check the tightness condition (d). Take $K_n = \bigcap_{m=n}^{\infty} R_m$. Then

$$\mathrm{Cap}(E \backslash K_n) \leqslant \sum_{m=n}^{\infty} \mathrm{Cap}(E \backslash R_m)$$

$$\leqslant \sum_{m=n}^{\infty} \left(\mathcal{E}(1_{E \backslash R_m}, 1_{E \backslash R_m}) + (1_{E \backslash R_m}, 1_{E \backslash R_m})_\mu \right)$$

$$= \sum_{m=n}^{\infty} \left(\Lambda([(E \backslash R_m) \times R_m] \cup [R_m \times (E \backslash R_m)]) + \mu(E \backslash R_m) \right).$$

The rightmost sum is finite by (d), and so we certainly have

$$\lim_{n \to \infty} \mathrm{Cap}(E \backslash K_n) = 0.$$

Finally, because constants belong to $\mathcal{D}(\mathcal{E})$, it follows from Theorem 1.6.3 of [72] that $\mathbf{X}$ is recurrent.

Remark 6.12. (i) Note that Example A.2 doesn't apply to give the closability of $\mathcal{E}$ unless $\nu_s = 0$.
(ii) Suppose that $\mathcal{S} \subseteq \mathcal{R}$ generates $\mathcal{R}$, then it suffices to check condition (b) just for $R \in \mathcal{S}$, as the following argument shows. We remarked in the proof that condition (b) was just the statement that $\mathcal{E}(1_R, 1_R) < \infty$ for all $R \in \mathcal{R}$. Note that 1_R for $R \in \mathcal{R}$ is a finite linear combination of functions of the form $f = \prod_{i=1}^{N} 1_{S_i}$ for $S_1, \dots, S_N \in \mathcal{S}$, and so it suffices to show that $\mathcal{E}(f, f) < \infty$ for such f. Observe that if $a_1, \dots, a_N \in \mathbb{R}$ and $b_1, \dots, b_N \in \mathbb{R}$ satisfy $|a_i| \leqslant 1$ and $|b_i| \leqslant 1$ for $1 \leqslant i \leqslant N$, then

$$\left| \prod_{i=1}^{N} a_i - \prod_{i=1}^{N} b_i \right| = \left| \sum_{i=1}^{N} \left(\prod_{j=1}^{i-1} a_j \right) (a_i - b_i) \left(\prod_{k=i+1}^{N} b_k \right) \right| \leqslant \sum_{i=1}^{N} |a_i - b_i|.$$

Therefore,

$$(f(y) - f(x))^2 = |f(y) - f(x)|$$

$$\leqslant \sum_{i=1}^{N} \left(1_{(E \backslash S_i) \times S_i}(x, y) + 1_{S_i \times (E \backslash S_i)}(x, y) \right),$$

and applying the assumption that (b) holds for all $R \in \mathcal{S}$ gives the result.
(iii) We emphasize that the elements of $\mathcal{D}(\mathcal{E})$ are elements of $L^2(E, \mu)$ and are thus equivalence classes of functions. It is clear from the above proof that if $f, g \in \mathcal{D}(\mathcal{E})$, then there are representatives $\hat{f}$ and $\hat{g}$ of the $L^2(E, \mu)$ equivalence classes of f and g such that

$$\mathcal{E}(f, g) = \iint (\hat{f}(y) - \hat{f}(x))(\hat{g}(y) - \hat{g}(x)) \, \Lambda(dx, dy).$$

Some care must be exercised here: it is clear that if $\nu_s \neq 0$, then we cannot substitute an arbitrary choice of representatives into the right–hand side to compute $\mathcal{E}(f, g)$.

(iv) The above proof appealed to Theorem A.8, which is Theorem 7.3.1 of
[72]. Although our state–space E is, in general, not locally compact, much
of the theory developed in [72] for the locally compact setting still applies
– see Remark A.9.

We present several examples of set-ups satisfying the conditions of the
Theorem 6.4 at the end of Section 6.4.

6.4 Bipartite Chains

Assume for this section that **X** is a bipartite chain.

Notation 6.13. *For a Borel set $B \subseteq E$, put $\sigma_B = \inf\{t > 0 : X_t \in B\}$ and $\tau_B = \inf\{t > 0 : X_t \notin B\}$.*

Proposition 6.14.

(i) *Consider $x \in E^o$. If $\int \kappa(x,z)\nu(dz) = 0$, then $\mathbb{P}^x\{\tau_{\{x\}} < \infty\} = 0$. Otherwise,*

$$\mathbb{P}^x\{\tau_{\{x\}} > t, X_{\tau_{\{x\}}} \in dy\} = \exp\left(-t \int \kappa(x,z)\nu(dz)\right) \frac{\kappa(x,y)\nu(dy)}{\int \kappa(x,z)\nu(dz)};$$

and, in particular, $\mathbb{P}^x\{X_{\tau_{\{x\}}} \in E^*\} = 1$.

(ii) *For q.e. $x \in E^*$, $\mathbb{P}^x\{X_t \in E^o\} = 1$ for Lebesgue almost all $t \geq 0$. In particular, $\mathbb{P}^x\{\sigma_{E^o} = 0\} = 1$ for q.e. $x \in E^*$.*

Proof. (i) Because each $x \in E^o$ is isolated, it follows from standard considerations that $\mathbb{P}\{\tau_{\{x\}} > t\} = \exp(-\alpha t)$, where

$$\mu(\{x\})\alpha = -\lim_{t\downarrow 0}\left(\frac{1}{t}(T_t - I)\mathbf{1}_x, \mathbf{1}_x\right)_\mu$$

$$= \mathcal{E}(\mathbf{1}_x, \mathbf{1}_x) = \mu(\{x\})\int \kappa(x,z)\nu(dz).$$

Observe for $f,g \in \mathcal{C}$ that $\mathcal{E}(f,g) = \iint(f(y) - f(x))(g(y) - g(x))\,J(dx,dy)$, where $J(dx,dy) = (1/2)[\Lambda(dx,dy) + \Lambda(dy,dx)]$ is the symmetrization of Λ. Note that J is a symmetric measure that assigns no mass to the diagonal of $E \times E$. This representation of $\mathcal{E}$ is the one familiar from the Beurling–Deny formula. The result now follows from Lemma 4.5.5 of [72].

(ii) This is immediate from the Markov property, Fubini and the observation $\mathbb{P}^\mu\{X_t \notin E^o\} = \mu(E^*) = 0$ for all $t \geq 0$. $\square$

Definition 6.15. *Define a subprobability kernel ξ on E by $\xi(x,B) = \mu \otimes \nu(\{(x',y) : \kappa(x,y) > 0, \kappa(x',y) > 0, x' \in B\})$. Note that $\xi(x,\cdot) \leq \mu$. Say that* **X** *is graphically irreducible if there exists $x_0 \in E^o$ such that for all $x \in E^o$ there exists $n \in \mathbb{N}$ for which $\xi^n(x_0, \{x\}) > 0$.*

Recall that a measure η is *invariant* for $\mathbf{X}$ if $\mathbb{P}^\eta\{X_t \in \cdot\} = \eta$ for all $t \geqslant 0$.

Proposition 6.16. *Suppose that $\mathbf{X}$ is graphically irreducible. Then μ is the unique invariant probability measure for $\mathbf{X}$ such that $\mu(E^o) = 1$. If γ is any other probability measure such that $\gamma(E^o) = 1$, then*

$$\lim_{t\to\infty} \sup_B |\mathbb{P}^\gamma\{X_t \in B\} - \mu(B)| = 0.$$

Proof. By standard coupling arguments, both claims will hold if we can show

$$\mathbb{P}^x\{\sigma_{\{y\}} < \infty\} = 1, \text{ for all } x, y \in E^o. \tag{6.8}$$

For (6.8) it suffices by Theorem 4.6.6 of [72] to check that the recurrent form $\mathcal{E}$ is irreducible in the sense of Section 1.6 of [72]. Furthermore, applying Theorem 1.6.1 of [72] (and the fact that $1 \in \mathcal{D}(\mathcal{E})$ with $\mathcal{E}(1,1) = 0$), it is certainly enough to establish that if B is any Borel set with $\mathbf{1}_B \in \mathcal{D}(\mathcal{E})$ and

$$0 = \mathcal{E}(\mathbf{1}_B, \mathbf{1}_B) + \mathcal{E}(\mathbf{1}_{E\setminus B}, \mathbf{1}_{E\setminus B}) = 2\mathcal{E}(\mathbf{1}_B, \mathbf{1}_B), \tag{6.9}$$

then $\mu(B)$ is either 0 or 1.

Suppose that (6.9) holds. By Remark 6.12(iii), there is a Borel function $\hat{f}$ with $\hat{f} = \mathbf{1}_B$, μ-a.e., such that

$$\begin{aligned} 0 &= \mathcal{E}(\mathbf{1}_B, \mathbf{1}_B) \\ &= \iint \left(\hat{f}(y) - \hat{f}(x)\right)^2 \Lambda(dx, dy) \\ &= \iint \left(\hat{f}(y) - \mathbf{1}_B(x)\right)^2 \Lambda(dx, dy). \end{aligned} \tag{6.10}$$

Suppose first that $x_0 \in B$, where x_0 is as in Definition 6.15. From (6.10),

$$\int \left(\hat{f}(y) - 1\right)^2 \kappa(x_0, y)\, \nu(dy) = 0,$$

and so $\nu(\{y : \hat{f} \neq 1, \kappa(x_0, y) > 0\}) = 0$. Therefore, again from (6.10), $\xi(x_0, \{x : \mathbf{1}_B(x) \neq 1\}) = 0$. That is, if $\xi(x_0, \{x\}) > 0$, then $x \in B$. Continuing in this way, we get that if $x \in E^o$ is such that $\xi^n(x_0, \{x\}) > 0$ for some n, then $x \in B$. Thus, $E^o \subseteq B$ and $\mu(B) = 1$. A similar argument shows that if $x_0 \notin B$, then $\mu(B) = 0$. □

Example 6.17. Suppose that we are in the setting of Example 6.1 with $E^o = \mathbb{N}$. Let μ be an arbitrary fully supported probability measure on $\mathbb{N}$ and put $\nu = \delta_\infty$. In order that the conditions of Theorem 6.4 hold we only need κ to satisfy $\sum_{x\in\mathbb{N}} \kappa(x, \infty)\mu(\{x\}) = \infty$. The conditions of Proposition 6.16 will hold if and only if $\kappa(x, \infty) > 0$ for all $x \in \mathbb{N}$.

Example 6.18. We recall the Dirichlet form for the wild chain. Here $E = \mathbf{T}^*$ from Example 6.10, μ is the PGW(1) distribution and ν is the distribution of a PGW(1) tree "conditioned to be infinite". A more concrete description of ν is the following. Each $y \in \mathbf{T}^*_\infty$ has a unique path $(u_0, u_1, u_2, \ldots)$ starting at the root. There is a bijection between $\mathbf{T}^*_\infty$ and $\mathbf{T}_{<\infty} \times \mathbf{T}_{<\infty} \times \ldots$ that is given by identifying $y \in \mathbf{T}^*_\infty$ with the sequence of finite trees $(y_0, y_1, y_2, \ldots)$, where y_i is the tree rooted at u_i in the forest obtained by deleting the edges of the path $(u_0, u_1, u_2, \ldots)$ – see Figure 6.2.

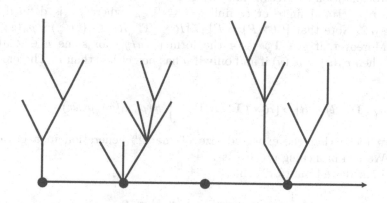

Fig. 6.2. The bijection between $\mathbf{T}^*_\infty$ and $\mathbf{T}_{<\infty} \times \mathbf{T}_{<\infty} \times \ldots$

The probability measure ν on $\mathbf{T}^*_\infty$ is the push–forward by this bijection of the probability measure $\mu \times \mu \times \ldots$ on $\mathbf{T}_{<\infty} \times \mathbf{T}_{<\infty} \times \ldots$

Rather than describe $\kappa(x, y)$ explicitly, it is more convenient (and equally satisfactory for our purposes) to describe the measures

$$q^\uparrow(x, dy) := \kappa(x, y)\, \nu(dy)$$

for each x and

$$q^\downarrow(y, dx) := \kappa(x, y)\, \mu(dx)$$

for each y. Given $x \in \mathbf{T}_{<\infty}$, $y \in \mathbf{T}^*_\infty$, and a vertex u of x, let $(x/u/y) \in \mathbf{T}^*_\infty$ denote the tree rooted at the root of x that is obtained by inserting a new edge from u to the root of y. Then

$$q^{\uparrow}(x, f) := \sum_{u \in x} \int f((x/u/y)) \, \nu(dy) \qquad (6.11)$$

for f a non-negative Borel function on $\mathbf{T}^*$.

For $y \in \mathbf{T}^*_\infty$ with infinite path from the root $(u_0, u_1, u_2, \ldots)$ and $i \in \mathbb{N}_0$, removing the edge (u_i, u_{i+1}) produces two trees, one finite rooted at u_0 and one infinite rooted at u_{i+1}. Let $k_i(y) \in \mathbf{T}_{<\infty}$ denote the finite tree. Then (6.11) is equivalent to

$$q^{\downarrow}(y, f) = \sum_{i \in \mathbb{N}_0} f(k_i(y)) \qquad (6.12)$$

for f a non-negative Borel function on $\mathbf{T}^*$.

Let us now check the conditions of Theorem 6.4. Condition (a) is obvious. Turning to condition (b), recall that any $R \in \mathcal{R}$ is of the form $\{x : \rho_n(x) \in B\}$ for some $n \in \mathbb{N}$ and finite or co-finite $B \subseteq \mathbf{T}_{\leq n}$, where ρ_n is defined in Example 6.2. Note that $[(\mathbf{T}^* \backslash R) \times R] \cup [R \times (\mathbf{T}^* \backslash R)] \subseteq \{(x, y) : \rho_n(x) \neq \rho_n(y)\}$. Moreover, if $y \in \mathbf{T}^*_\infty$ is of the form $(x/u/y')$ for some $u \in x$ and $y' \in \mathbf{T}^*_\infty$, then $\rho_n(x) \neq \rho_n(y)$ if and only if u has height less than n. Therefore, by (6.11),

$$\Lambda([(\mathbf{T}^* \backslash R) \times R] \cup [R \times (\mathbf{T}^* \backslash R)]) \leq \int \#(\rho_{n-1}(x)) \, \mu(dx) = n,$$

where we recall that the expected size of the kth generation in a critical Galton–Watson branching process is 1.

It is immediate from (6.12) that

$$\int \kappa(x, y) \, \mu(dx) = q^{\downarrow}(y, 1) = \infty$$

for $\nu = \nu_s$ almost every y, and so condition (c) holds.

Finally, consider condition (d). Put $S_{n,c} := \{x : \#(\rho_n(x)) \leq c\}$. We will take $R_n = S_{n,c_n}$ for some sequence of constants $(c_n)_{n \in \mathbb{N}}$. Note that $\bigcap_{m=n}^{\infty} S_{m,c_m}$ is compact for all n, whatever the choice of $(c_n)_{n \in \mathbb{N}}$. By choosing c_n large enough, we can certainly make $\mu(\mathbf{T}^* \backslash S_{n,c_n}) \leq 2^{-n}$. From the argument for part (b) we know that $[(\mathbf{T}^* \backslash S_{n,c}) \times S_{n,c}] \cup [S_{n,c} \times (\mathbf{T}^* \backslash S_{n,c})] = S_{n,c} \times (\mathbf{T}^* \backslash S_{n,c})$ is contained in the set $\{(x, y) : \rho_n(x) \neq \rho_n(y)\}$ that has finite Λ measure. Of course, $\lim_{c \to \infty} \mathbf{T}^* \backslash S_{n,c} = \varnothing$. Therefore, by dominated convergence, $\lim_{c \to \infty} \Lambda([(\mathbf{T}^* \backslash S_{n,c}) \times S_{n,c}] \cup [S_{n,c} \times (\mathbf{T}^* \backslash S_{n,c})]) = 0$, and by choosing c_n large enough we can make $\Lambda([(\mathbf{T}^* \backslash S_{n,c_n}) \times S_{n,c_n}] \cup [S_{n,c_n} \times (\mathbf{T}^* \backslash S_{n,c_n})]) \leq 2^{-n}$.

It is obvious that the extra bipartite chain conditions hold with $E^o = \mathbf{T}_{<\infty}$. The condition of Proposition 6.16 also holds. More specifically, we can take x_0 in Definition 6.15 to be the trivial tree consisting of only a root. By (6.11) and (6.12), the measure $\xi^n(x_0, \cdot)$ assigns positive mass to every tree $x \in \mathbf{T}_{<\infty}$ with at most n children in the first generation (that is, $x \in \mathbf{T}_{<\infty}$ such that $\#(\rho_1(x)) \leq n + 1$), and so $\mathbf{X}$ is indeed graphically irreducible.

Example 6.19. Suppose that we are in the setting of Example 6.8 with $\#S < \infty$ (so that E is compact) and E^o the set $\{x : \zeta(x) < \infty\}$, as above. Note that $E^* = S^{\mathbb{N}_0}$. Fix a probability measure P on S with full support, an $S \times S$ stochastic matrix Q with positive entries and a probability measure R on $\mathbb{N}_0$. Define a probability measure μ on E^o by $\mu(\{x : \zeta(x) = n, x_0 = s_0, \ldots, x_{n-1} = s_{n-1}\}) = R(n)P(s_0)Q(s_0, s_1)\ldots Q(s_{n-2}, s_{n-1})$. In other words, μ is the law of a Markov chain with initial distribution P and transition matrix Q killed at an independent time with distribution R. Define ν on E^* by $\nu(\{s_0\} \times \cdots \times \{s_n\} \times S \times S \times \ldots) = P(s_0)Q(s_0, s_1)\ldots Q(s_{n-1}, s_n)$. Thus, ν is the law of the unkilled chain with initial distribution P and transition matrix Q. Define $\kappa(x, y)$ for $x \in E^o$ and $y \in E^*$ by $\kappa(x, y) = K(\zeta(x))\mathbf{1}_{x \leqslant y}$ for some sequence of non-negative constants $K(n)$, $n \in \mathbb{N}_0$.

In order that the conditions of Theorem 6.4 hold, we only need K to satisfy $\sum_{x \leqslant y} K(\zeta(x))\mu(\{x\}) = \infty$ for ν-a.e. $y \in E^*$. For example, if $q_* = \min_{s, s'} Q(s, s')$, then it suffices that $\sum_{n \in \mathbb{N}_0} K(n)R(n)q_*^n = \infty$. In particular, if ν is the law of a sequence of i.i.d. uniform draws from S (so that $P(s) = S(s, s') = (\#S)^{-1}$ for all $s, s' \in S$), then we require $\sum_{n \in \mathbb{N}_0} K(n)R(n)(\#S)^{-n} = \infty$.

In general, **X** will be graphically irreducible with $x_0 = (\partial, \partial, \ldots)$ (and, therefore, the condition of Proposition 6.16 holds) if $K(n) > 0$ for all $n \in \mathbb{N}_0$.

6.5 Quotient Processes

Return to the general set-up of Theorem 6.4. Suppose that $\mathcal{R}'$ is a subalgebra of $\mathcal{R}$ and write $\mathcal{C}'$ for the subalgebra of $\mathcal{C}$ generated by the indicator functions of sets in $\mathcal{R}'$. We can define an equivalence relation on E by declaring that x and y are equivalent if $f(x) = f(y)$ for all $f \in \mathcal{C}'$. Let $\bar{E}$ denote the corresponding quotient space equipped with the quotient topology and denote by $\pi : E \to \bar{E}$ the quotient map. It is not hard to check that $\bar{E}$ is a Lusin space and that the algebra $\bar{\mathcal{R}} := \{\pi R : R \in \mathcal{R}'\}$ consists of simultaneously closed and open sets and is a base for the topology of $\bar{E}$. Write $\bar{\mathcal{C}}$ for the algebra generated by the indicator functions of sets in $\bar{\mathcal{R}}$. Note that $\mathcal{C}' = \{\bar{f} \circ \pi : \bar{f} \in \bar{\mathcal{C}}\}$.

Proposition 6.20. *Suppose that the following hold:*

(a) $\mu = \nu$;

(b) there exists a Borel function $\bar{\kappa} : \bar{E} \times \bar{E} \to \mathbb{R}_+$ such that $\kappa(x, y) = \bar{\kappa}(\pi x, \pi y)$ for $\pi x \neq \pi y$;

(c) $\bar{E}$ is compact;

(d) $\mu_{\mathcal{R}'}[f] := \mu[f|\sigma(\mathcal{R}')] = \mu[f|\sigma(\pi)]$ has a version in $\mathcal{C}'$ for all $f \in \mathcal{C}$.

Then the hypotheses of Theorem 6.4 hold with E, $\mathcal{R}$, $\mathcal{C}$, μ, ν, κ replaced by $\bar{E}$, $\bar{\mathcal{R}}$, $\bar{\mathcal{C}}$, $\bar{\mu}$, $\bar{\nu}$, $\bar{\kappa}$, where $\bar{\mu} = \bar{\nu}$ is the push-forward of $\mu = \nu$ by π. Moreover, if $(\bar{\mathcal{E}}, \mathcal{D}(\bar{\mathcal{E}}))$ denotes the resulting Dirichlet form, then $\pi \circ \mathbf{X}$ is a $\bar{\mu}$-symmetric Hunt process with Dirichlet form $(\bar{\mathcal{E}}, \mathcal{D}(\bar{\mathcal{E}}))$.

Proof. It is clear that the hypotheses of Theorem 6.4 hold with $E, \mathcal{R}, \mathcal{C}, \mu, \nu, \kappa$ replaced by $\bar{E}, \bar{\mathcal{R}}, \bar{\mathcal{C}}, \bar{\mu}, \bar{\nu}, \bar{\kappa}$.

Let $(\bar{T}_t)_{t \geqslant 0}$ denote the semigroup on $L^2(\bar{E}, \bar{\mu})$ corresponding to $\bar{\mathcal{E}}$. The proof $\pi \circ \mathbf{X}$ is a $\bar{\mu}$-symmetric Hunt process with Dirichlet form $(\bar{\mathcal{E}}, \mathcal{D}(\bar{\mathcal{E}}))$ will be fairly straightforward once we establish that $T_t(\bar{f} \circ \pi) = (\bar{T}_t \bar{f}) \circ \pi$ for all $t \geqslant 0$ and $\bar{f} \in L^2(\bar{E}, \bar{\mu})$ (see Theorem 13.5 of [128] for a proof that this suffices for $\pi \circ \mathbf{X}$ to be a Hunt process – the proof that $\pi \circ \mathbf{X}$ is $\bar{\mu}$-symmetric and the identification of the associated Dirichlet form are then easy). Equivalently, writing $(G_\alpha)_{\alpha>0}$ and $(\bar{G}_\alpha)_{\alpha>0}$ for the resolvents corresponding to $(T_t)_{t \geqslant 0}$ and $(\bar{T}_t)_{t \geqslant 0}$, we need to establish that $G_\alpha(\bar{f} \circ \pi) = (\bar{G}_\alpha \bar{f}) \circ \pi$ for all $\alpha > 0$ and $\bar{f} \in L^2(\bar{E}, \bar{\mu})$. This is further equivalent to establishing that $(\bar{G}_\alpha \bar{f}) \circ \pi \in \mathcal{D}(\mathcal{E})$ and $\mathcal{E}((\bar{G}_\alpha \bar{f}) \circ \pi, g) + \alpha((\bar{G}_\alpha \bar{f}) \circ \pi, g)_\mu = (\bar{f} \circ \pi, g)_\mu$ for all $g \in \mathcal{C}$ – see Equation (1.3.7) of [72].

Fix $\bar{f} \in L^2(\bar{E}, \bar{\mu})$ and $g \in \mathcal{C}$. By assumption, $\mu_{\mathcal{R}'}[g] = \bar{g} \circ \pi$ for some $\bar{g} \in \bar{\mathcal{C}}$. Also, it is fairly immediate from the definition of $\bar{E}$ that $\bar{h} \in \mathcal{D}(\bar{\mathcal{E}})$ if and only if $\bar{h} \circ \pi \in \mathcal{D}(\mathcal{E})$, and that $\bar{\mathcal{E}}(\bar{h}, \bar{h}) = \mathcal{E}(\bar{h} \circ \pi, \bar{h} \circ \pi)$. Hence, by Remark 6.12(iii),

$$\mathcal{E}(\bar{h} \circ \pi, g) = \iint (\bar{h} \circ \pi(y) - \bar{h} \circ \pi(x)) \, (g(y) - g(x)) \, \Lambda(dx, dy)$$

$$= \iint_{\{(x,y):\pi x \neq \pi y\}} (\bar{h} \circ \pi(y) - \bar{h} \circ \pi(x)) \, (g(y) - g(x)) \, \Lambda(dx, dy)$$

$$= \iint (\bar{h} \circ \pi(y) - \bar{h} \circ \pi(x)) \, (g(y) - g(x)) \, \bar{\kappa}(\pi x, \pi y) \, \mu(dx) \, \mu(dy)$$

$$= \iint (\bar{h} \circ \pi(y) - \bar{h} \circ \pi(x)) \, (\mu_{\mathcal{R}'}[g](y) - \mu_{\mathcal{R}'}[g](x)) \, \bar{\kappa}(\pi x, \pi y) \, \mu(dx) \, \mu(dy)$$

$$= \iint (\bar{h} \circ \pi(y) - \bar{h} \circ \pi(x)) \, (\bar{g} \circ \pi(y) - \bar{g} \circ \pi(x)) \, \bar{\kappa}(\pi x, \pi y) \, \mu(dx) \, \mu(dy)$$

$$= \iint (\bar{h}(w) - \bar{h}(v)) \, (\bar{g}(w) - \bar{g}(v)) \, \bar{\kappa}(v, w) \, \bar{\mu}(dv) \, \bar{\mu}(dw)$$

$$= \bar{\mathcal{E}}(\bar{h}, \bar{g}).$$

Of course,

$$(\bar{h} \circ \pi, g)_\mu = (\bar{h} \circ \pi, \bar{g} \circ \pi)_\mu = (\bar{h}, \bar{g})_{\bar{\mu}}.$$

Therefore,

$$\mathcal{E}((\bar{G}_\alpha \bar{f}) \circ \pi, g) + \alpha((\bar{G}_\alpha \bar{f}) \circ \pi, g)_\mu = \bar{\mathcal{E}}(\bar{G}_\alpha \bar{f}, \bar{g}) + \alpha(\bar{G}_\alpha \bar{f}, \bar{g})_{\bar{\mu}}$$
$$= (\bar{f}, \bar{g})_{\bar{\mu}} = (\bar{f} \circ \pi, \bar{g} \circ \pi)_\mu = (\bar{f} \circ \pi, g)_\mu,$$

as required. □

We will see an application of Proposition 6.20 at the end of Section 6.7.

6.6 Additive Functionals

We are still in the general setting of Theorem 6.4.

Proposition 6.21. *The probability measure ν assigns no mass to sets of zero capacity, and there is a positive continuous additive functional $(A_t)_{t \geq 0}$ with Revuz measure ν.*

Proof. The reference measure μ assigns no mass to sets of zero capacity, so it suffices to show that ν_s assigns no mass to sets of zero capacity. For $M > 0$ put $G_M := \{ y : \int [\kappa(x,y) \wedge M] \, \mu(dx) \geq 1 \}$ and define a subprobability measure ν_s^M by $\nu_s^M := \nu_s(\cdot \cap G_M)$. By (c) of Theorem 6.4, $\nu_s(E \backslash \bigcup_M G_M) = 0$, and so it suffices to show for each M that ν_s^M assigns no mass to sets of zero capacity.

Observe for $f \in \mathcal{C}$ that

$$\left(\int |f(y)| \, \nu_s^M(dy) \right)^2 \leq \int f^2(y) \, \nu_s^M(dy) \leq \iint f^2(y) \, \Lambda^M(dx, dy)$$

$$\leq 2 \left(\iint (f(y) - f(x))^2 \, \Lambda^M(dx, dy) + \iint f^2(x) \, \Lambda^M(dx, dy) \right)$$

$$\leq 2(1 \vee M) \left(\mathcal{E}(f,f) + (f,f)_\mu \right).$$

The development leading to Lemma 2.2.3 of [72] can now be followed to show that for all Borel sets B we have $\nu_s^M(B) \leq C_M \mathrm{Cap}(B)^{1/2}$ for a suitable constant C_M (the argument in [72] is in a locally compact setting, but it carries over without difficulty to our context).

The existence and uniqueness of $(A_t)_{t \geq 0}$ follows from Theorem 5.1.4 of [72]. □

Remark 6.22. In the bipartite chain case, the distribution under $\mathbb{P}^\mu$ of X_ζ, where $\zeta := \tau_{\{X_0\}}$, is mutually absolutely continuous with respect to ν, and Proposition 6.21 is obvious.

6.7 Bipartite Chains on the Boundary

Return to the bipartite chain setting. Following the construction in Section 6.2 of [72], let $\mathbf{Y}$ denote the process $\mathbf{X}$ time–changed according to the positive continuous additive functional A. That is, $Y_t = X_{\gamma_t}$ where $\gamma_t = \inf\{s > 0 : A_s > t\}$. Write $\tilde{E}$ for the support of A. We have $\tilde{E} \subseteq \breve{E} := \mathrm{supp}\, \nu \subseteq E^*$ and $\nu(\breve{E} \backslash \tilde{E}) = 0$.

Let $\breve{\mathcal{R}} = \{R \cap \breve{E} : R \in \mathcal{R}\}$ and put $\breve{\mathcal{C}} = \{f_{|\breve{E}} : f \in \mathcal{C}\}$. Note that $\breve{\mathcal{C}}$ is also the algebra generated by $\breve{\mathcal{R}}$.

Theorem 6.23. *The process $\mathbf{Y}$ is a recurrent ν-symmetric Hunt process with state-space $\breve{E}$ and Dirichlet form given by the closure of the form $\breve{\mathcal{E}}$ on $\breve{\mathcal{C}}$ defined by*

$$\breve{\mathcal{E}}(f,g) = \iint (f(y) - f(z))(g(y) - g(z)) \, \breve{\kappa}(y,z) \, \nu(dy) \, \nu(dz), \quad f, g \in \breve{\mathcal{C}},$$

where

$$\check{\kappa}(y, z) = \int \kappa(x, y) \frac{\kappa(x, z)}{\int \kappa(x, w) \, \nu(dw)} \, \mu(dx)$$

(with the convention $0/0 = 0$).

Proof. By Theorem A.2.6 and Theorem 4.1.3 of [72],

$$\mathbb{P}^y\{\sigma_{\check{E}} = 0\} = 1 \text{ for q.e. } y \in \check{E}.$$

Hence, for q.e. $y \in \check{E}$ we have $\lim_{\epsilon \downarrow 0} \inf\{t > \epsilon : X_t \in \check{E}\} = 0$, $\mathbb{P}^y$-a.s. More-over, it follows from parts (i) and (ii) of Proposition 6.14 and the observation $\nu(\check{E} \backslash \tilde{E}) = 0$ that for q.e. $y \in \check{E}$ we have $\inf\{t > \epsilon : X_t \in \check{E}\} = \inf\{t > \epsilon : X_t \in \tilde{E}\}$ for all $\epsilon > 0$, $\mathbb{P}^y$-a.s. Combining this with Proposition 6.21 gives

$$\mathbb{P}^y\{\sigma_{\tilde{E}} = 0\} = 1 \text{ for q.e. and } \nu\text{-a.e. } y \in \check{E}.$$

Define $H_{\tilde{E}} f(x) := \mathbb{P}^x[f(X_{\sigma_{\tilde{E}}})]$ for f a bounded Borel function on E. It follows from part (i) of Proposition 6.14 and what we have just observed that

$$H_{\tilde{E}} f(x) = \frac{\int f(y)\kappa(x, y) \, \nu(dy)}{\int \kappa(x, y) \, \nu(dy)}, \text{ for } \mu\text{-a.e. } x$$

and

$$H_{\tilde{E}} f(x) = f(x), \text{ for } \nu\text{-a.e. } x.$$

The result now follows by applying Theorem 6.2.1 of [72].   □

Example 6.24. Suppose that we are in the setting of Example 6.19. For $y, z \in E^* = S^{\mathbb{N}_0}$, $y \neq z$, define $\delta(y, z) = \inf\{n : y_n \neq z_n\}$. Note that $\int \kappa(x, w) \, \nu(dw) = K(\zeta(x))\nu(\{w : x \leqslant w\}) = K(\zeta(x))\mu(\{x\})/R(\zeta(x))$ for $x \in E^o$ and so

$$\check{\kappa}(y, z) = \sum_{n \leqslant \delta(y, z)} K(n)R(n). \tag{6.13}$$

We will now apply the results of Section 6.5 with $E, \mathbf{X}, \mu, \mathcal{E}$ replaced by $\check{E} = E^* = S^{\mathbb{N}_0}, \mathbf{Y}, \nu, \check{\mathcal{E}}$. Fix $N \in \mathbb{N}_0$ and let $\mathcal{R}'$ be the algebra of subsets of $S^{\mathbb{N}_0}$ of the form $B_0 \times \cdots \times B_N \times S \times S \times \ldots$. We can identify the quotient space $\bar{E}$ with S^{N+1} and the quotient map π with the map $(y_0, y_1, \ldots) \mapsto (y_0, \ldots y_N)$. Then we can identify $\bar{\mu}$, which we emphasise is now the push–forward ν by π, with the measure that assigns mass $P(s_0)Q(s_0, s_1) \ldots Q(s_{N-1}, s_N)$ to $(s_0, \ldots s_N)$. Note that $\pi y \neq \pi z$ for $y, z \in S^{\mathbb{N}_0}$ is equivalent to $\delta(y, z) \leqslant N$, and it is immediate from (6.13) that Proposition 6.20 applies and $\pi \circ \mathbf{Y}$ is a $\bar{\mu}$-symmetric Markov chain on the finite state-space S^{N+1}. In terms of jump rates, $\pi \circ \mathbf{Y}$ jumps from $\bar{y}$ to $\bar{z} \neq \bar{y}$ at rate $(\sum_{n \leqslant \delta(\bar{y}, \bar{z})} K(n)R(n))\bar{\mu}(\{\bar{z}\})$, where $\delta(\bar{y}, \bar{z})$ is defined in the obvious way.

As a particular example of this construction, consider the case when $\#S = p^c$ for some prime p and integer $c \geqslant 1$. We can identify $S^{\mathbb{N}_0}$ (as a set) with the ring of integers $\mathbb{D}$ of a local field $\mathbb{K}$ as in Example 6.9. If we take $P(s) = Q(s, s') = p^{-c}$ for all $s, s' \in S$, then we can identify ν with the normalised Haar measure on $\mathbb{D}$. It is clear that $\mathbf{Y}$ is a Lévy process on $\mathbb{D}$ with "spherically symmetric" Lévy measure $\phi(|y|)\,\nu(dy)$, where $\phi(p^{-cn}) = \sum_{\ell=0}^{n} K(\ell)R(\ell)$. The condition $\sum_{n \in \mathbb{N}_0} K(n)R(n)p^{-cn} = \infty$ of Example 6.19 is equivalent to $\int_{\mathbb{D}} \phi(|y|)\,\nu(dy) = \infty$. Conversely, any Lévy process on $\mathbb{D}$ with Lévy measure of the form $\psi(|y|)\,\nu(dy)$ with ψ non-increasing and $\int_{\mathbb{D}} \psi(|y|)\,\nu(dy) = \infty$ can be produced by this construction (Lévy processes on $\mathbb{D}$ are completely characterised by their Lévy measures – there is no analogue of the drift or Gaussian components of the Euclidean case, see [59]). The latter condition is equivalent to the paths of the process almost surely not being step–functions, that is, to the times at which jumps occur being almost surely dense. When $\psi(|y|) = a|y|^{-(\alpha+1)}$ for some $a > 0$ and $0 < \alpha < \infty$, the resultant process is analogous to a symmetric stable process. Lévy processes on local fields and totally disconnected Abelian groups in general are considered in [59] and the special case of the p-adic numbers has been considered by a number of authors – see Chapter 1 for a discussion.

Diffusions on a $\mathbb{R}$-Tree without Leaves: Snakes and Spiders

7.1 Background

Let (T, d) be a $\mathbb{R}$-tree without ends as in Section 3.4. Suppose that that there is a σ-finite Borel measure μ on the set on $\mathbf{E}_+$ of ends at $+\infty$ such that $0 < \mu(B) < \infty$ for every ball B in the metric δ. In particular, the support of μ is all of E_+.

The existence of such a measure μ is a more restrictive assumption on T than it might first appear. Let $\bar{\mu}$ be a finite measure on E_+ that is equivalent to μ. Recall from (3.4) that T_t, $t \in \mathbb{R}$, is the set of points in T with height t. As we remarked in Section 3.4.2, the set $\{\zeta \in E_+ : \zeta|t = x\}$ is a ball in E_+ for each $x \in T_t$ and two such balls are disjoint. Because the $\bar{\mu}$ measure of each such ball is non–zero, the set T_t is necessarily countable. Hence, by observations made in Section 3.4.2, both the complete metric spaces T and E_+ are separable, and, therefore, Lusin.

We will be interested in the T–valued process X that evolves in the following manner. The real–valued process H, where $H_t = h(X_t)$, evolves as a standard Brownian motion. For small $\epsilon > 0$ the conditional probability of the event $\{X_{t+\epsilon} \in C\}$ given X_t and H is approximately

$$\frac{\mu \{y : y | H_{t+\epsilon} \in C, \, y | H_t = X_t\}}{\mu \{y : y | H_t = X_t\}}.$$

In particular, if $H_{t+\epsilon} < H_t$, then $X_{t+\epsilon}$ is approximately $X_t | H_{t+\epsilon}$. An intuitive description of these dynamics is given in Figure 7.1.

This evolution is reminiscent of Le Gall's *Brownian snake* process – see, for example, [97, 98, 99, 100] – with the difference that the "height" process H is a Brownian motion here rather than a reflected Brownian motion and the role of Wiener measure on $C(\mathbb{R}_+, \mathbb{R}^d)$ in the snake construction is played here by μ.

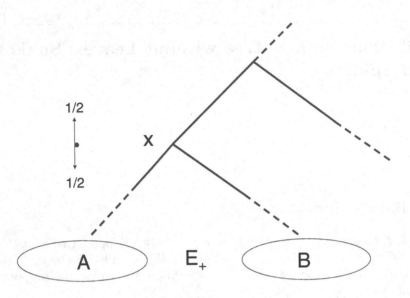

Fig. 7.1. A heuristic description of the dynamics of X. When X is at position x it makes an infinitesimal move up or down with equal probability. Conditional on X moving down, it takes the branch leading to the set of ends A with probability $\mu(A)/(\mu(A) + \mu(B))$ and the branch leading to the set of ends B with probability $\mu(B)/(\mu(A) + \mu(B))$.

7.2 Construction of the Diffusion Process

For $x \in T$ and real numbers $b < c$ with $b < h(x)$, define a probability measure $\mu(x, b, c; \cdot)$ on T by

$$\mu(x, b, c; A) := \frac{\mu\{\xi \in E_+ : \xi | c \in A, \, \xi | b = x | b\}}{\mu\{\xi \in E_+ : \xi | b = x | b\}}.$$

– see Figure 7.2.

Let (B_t, P^a) be a standard (real–valued) Brownian motion. Write $m_t := \inf_{0 \leqslant s \leqslant t} B_s$. Recall that the pair (m_t, B_t) has joint density

$$\phi_{a,t}(b, c) := \sqrt{\frac{2}{\pi}} \frac{c - 2b + a}{t^{3/2}} \exp\left(-\frac{(c - 2b + a)^2}{2t}\right), \quad b < a \wedge c,$$

under P^a – see, for example, Corollary 30 in Chapter 1 of [70].

Theorem 7.1. *There is a Markov semigroup $(P_t)_{t \geqslant 0}$ on T defined by*

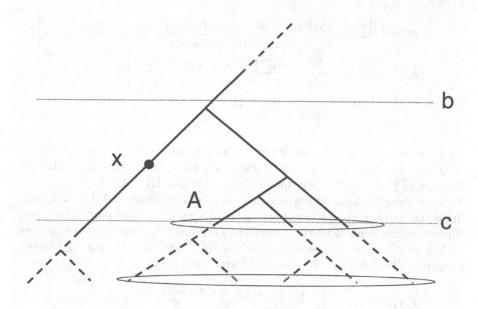

Fig. 7.2. The measure $\mu(x, b, c; \cdot)$ is supported on the set $\{y \in T : h(y) = c, y|b = x|b\}$, and the mass it assigns to the set A is the normalized μ mass of the shaded subset of E_+.

$$P_t f(x) := P^{h(x)} \left[\mu(x, m_t, B_t; f) \right].$$

Furthermore, there is a strong Markov process $(X_t, \mathbb{P}^x)$ on T with continuous sample paths and semigroup $(P_t)_{t \geq 0}$.

Proof. The proof of the semigroup property of $(P_t)_{t \geq 0}$ is immediate from the Markov property of Brownian motion and the readily checked observation that for $x, x' \in T$, $b < c$, $b < h(x)$, and $b' < c \wedge c'$ we have

$$\int \mu(x', b', c'; A) \, \mu(x, b, c; dx') = \mu(x, b \wedge b', c'; A).$$

By Kolmogorov's extension theorem, there is a Markov process $(X_t, \mathbb{P}^x)$ on T with semigroup $(P_t)_{t \geq 0}$. In order to show that a version of X can be chosen with continuous sample paths, it suffices because (T, d) is complete and separable to check Kolmogorov's continuity criterion. Because of the Markov property of X, it further suffices to observe for $\alpha > 0$ that, by definition of $(P_t)_{t \geq 0}$,

$$\mathbb{P}^x \left[d(x, X_t)^\alpha \right]$$

$$= P^{h(x)} \left[\frac{\int [h(x) + h(\xi|B_t) - 2h(x \wedge (\xi|B_t))]^\alpha \mathbf{1}\{\xi|m_t = x|m_t\} \mu(d\xi)}{\mu\{\xi \in E_+ : \xi|m_t = x|m_t\}} \right]$$

$$\leqslant P^{h(x)} \left[\frac{\int [h(x) + B_t - 2m_t]^\alpha \mathbf{1}\{\xi|m_t = x|m_t\} \mu(d\xi)}{\mu\{\xi \in E_+ : \xi|m_t = x|m_t\}} \right]$$

$$\leqslant C P^{h(x)} \left[|h(x) - m_t|^\alpha + |m_t - B_t|^\alpha \right]$$

$$\leqslant C' t^{\alpha/2}$$

for some constants C, C' that depend on α but not on $x \in T$.

The claim that X is strong Markov will follow if we can show that P_t maps $bC(T)$ into itself – see, for example, Sections III.8, III.9 of [120]. It is assumed there that the underlying space is locally compact and the semigroup maps the space of continuous functions that vanish at infinity into itself, but this stronger assumption is only needed to establish the existence of a process with càdlàg sample paths and plays no role in the proof of the strong Markov property). By definition, for $f \in b\mathcal{B}(T)$ and $t > 0$

$$P_t f(x) = \int_{-\infty}^{h(x)} \int_b^\infty \frac{\int f(\xi|c) \mathbf{1}\{\xi|b = x|b\} \mu(d\xi)}{\mu\{\xi \in E_+ : \xi|b = x|b\}}$$
$$\times \sqrt{\frac{2}{\pi}} \frac{c - 2b + h(x)}{t^{3/2}} \exp\left(-\frac{(c - 2b + h(x))^2}{2t}\right) dc\, db$$

for $t > 0$. The right–hand side can be written as $\int_{-\infty}^\infty \int_{-\infty}^\infty F_{f,x}(b, c)\, dc\, db$ for a certain function $F_{f,x}$. Recall from (3.2) that $|h(x) - h(x')| \leqslant d(x, x')$. Also, if $b < h(x)$, then $x'|b = x|b$ for x' such that $d(x, x') \leqslant h(x) - b$. Therefore, $\lim_{x' \to x} F_{f,x'}(b, c) = F_{f,x}(b, c)$ except possibly at $b = h(x)$. Moreover, if $\sup_x |f(x)| \leqslant C$, then $|F_{f,x}(b, c)| \leqslant C F_{1,x}(b, c)$. Because

$$\lim_{x' \to x} \int_{-\infty}^\infty \int_{-\infty}^\infty F_{1,x'}(b, c)\, dc\, db = \lim_{x' \to x} 1 = 1 = \int_{-\infty}^\infty \int_{-\infty}^\infty F_{1,x}(b, c)\, dc\, db,$$

a standard generalization of the dominated convergence theorem – see, for example, Proposition 18 in Chapter 11 of [121] – shows that if $f \in b\mathcal{B}(T)$, then $P_t f \in bC(T)$ for $t > 0$. □

7.3 Symmetry and the Dirichlet Form

Write λ for Lebesgue measure on $\mathbb{R}$. Consider the measure ν that is obtained by pushing forward the measure $\mu \otimes \lambda$ on $E_+ \times \mathbb{R}$ with the map $(\xi, a) \mapsto \xi|a$ – see Figure 7.3. Note that for $x \in T$ with $h(x) = h^*$ and $\epsilon > 0$ we have

$$\nu\{y \in T : d(x, y) \leqslant \epsilon\}$$
$$\leqslant \nu\{y \in T : y|(h^* - \epsilon) = x|(h^* - \epsilon), h^* - \epsilon \leqslant h(y) \leqslant h^* + \epsilon\}$$
$$\leqslant 2\epsilon \mu\{\xi \in E_+ : \xi|(h^* - \epsilon) = x|(h^* - \epsilon)\}.$$

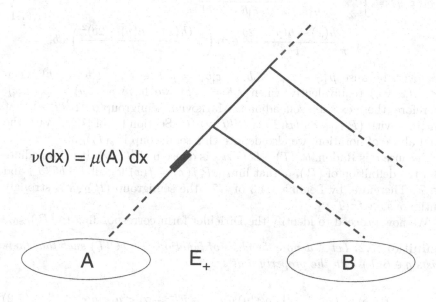

$\nu(dx) = \mu(A)\, dx$

Fig. 7.3. The definition of the measure ν on T in terms of the measure μ on E_+.

That is, ν assigns finite mass to balls in T and, in particular, is Radon.

We begin by showing that each operator P_t, $t > 0$, can be continuously extended from $b\mathcal{B}(T) \cap L^2(T, \nu)$ to $L^2(T, \nu)$ and that the resulting semigroup is a strongly continuous, self–adjoint, Markovian semigroup on $L^2(T, \nu)$.

Observe that if $f \in b\mathcal{B}(T)$, then

$$
\begin{aligned}
P_t f(x) &= \int_{E_+} \int_{\mathbb{R}} \int_{-\infty}^{h(x) \wedge c} \frac{f(\xi|c)\mathbf{1}\{\xi|b = x|b\}}{\mu\{\xi \in E_+ : \xi|b = x|b\}} \phi_{h(x),t}(b,c)\, db\, dc\, \mu(d\xi) \\
&= \int_T f(y) \int_{-\infty}^{h(x) \wedge h(y)} \frac{\mathbf{1}\{x|b = y|b\}}{\mu\{\xi \in E_+ : \xi|b = x|b\}} \\
&\quad \times \sqrt{\frac{2}{\pi}} \frac{h(x) + h(y) - 2b}{t^{3/2}} \exp\left(-\frac{(h(x) + h(y) - 2b)^2}{2t}\right) db\, \nu(dy) \\
&= \int_T f(y) \int_{-\infty}^{h(x \wedge y)} \frac{1}{\mu\{\xi \in E_+ : \xi|b = x|b\}} \\
&\quad \times \sqrt{\frac{2}{\pi}} \frac{h(x) + h(y) - 2b}{t^{3/2}} \exp\left(-\frac{(h(x) + h(y) - 2b)^2}{2t}\right) db\, \nu(dy)
\end{aligned}
$$

for $t > 0$. Consequently, $P_t f(x) = \int_T p_t(x,y) f(y)\, \nu(dy)$ for the jointly continuous, everywhere positive transition density

$$p_t(x,y) := \int_{-\infty}^{h(x \wedge y)} \frac{1}{\mu\{\xi \in E_+ : \xi|b = x|b\}}$$
$$\times \sqrt{\frac{2}{\pi}} \frac{h(x) + h(y) - 2b}{t^{3/2}} \exp\left(-\frac{(h(x) + h(y) - 2b)^2}{2t}\right) db. \qquad (7.1)$$

Moreover, because $\mu\{\xi \in E_+ : \xi|b = x|b\} = \mu\{\xi \in E_+ : \xi|b = y|b\}$ when $b \leqslant h(x \wedge y)$ (equivalently, when $x|b = y|b$), we have $p_t(x,y) = p_t(y,x)$. Therefore, there exists a self-adjoint, Markovian semigroup on $L^2(T, \nu)$ that coincides with $(P_t)_{t \geqslant 0}$ on $b\mathcal{B}(T) \cap L^2(T, \nu)$ (cf. Section 1.4 of [72]). With the usual abuse of notation, we also denote this semigroup by $(P_t)_{t \geqslant 0}$.

Because ν is Radon, $bC(T) \cap L^1(T, \nu)$ is dense in $L^2(T, \nu)$. It is immediate from the definition of $(P_t)_{t \geqslant 0}$ that $\lim_{t \downarrow 0} P_t f(x) = f(x)$ for all $f \in bC(T)$ and $x \in T$. Therefore, by Lemma 1.4.3 of [72], the semigroup $(P_t)_{t \geqslant 0}$ is strongly continuous on $L^2(T, \nu)$.

We now proceed to identify the Dirichlet form corresponding to $(P_t)_{t \geqslant 0}$.

Definition 7.2. *Let $\mathcal{A}$ denote the class of functions $f \in bC(T)$ such that there exists $g \in \mathcal{B}(T)$ with the property that*

$$f(\xi|b) - f(\xi|a) = \int_a^b g(\xi|u) \, du, \quad \xi \in E_+, \ -\infty < a < b < \infty. \qquad (7.2)$$

Note for $\xi \in E_+$ that if $A \in \mathcal{B}(T)$ with $A \subseteq [a,b]$, where $-\infty < a < b < \infty$, then

$$\mu\{\zeta \in E_+ : \zeta|b = \xi|b\} \lambda(A) \leqslant \nu\{\xi|u : u \in A\}$$
$$\leqslant \mu\{\zeta \in E_+ : \zeta|a = \xi|a\} \lambda(A).$$

Therefore, the function g in (7.2) is unique up to ν-null sets, and (with the usual convention of using function notation to denote equivalence classes of functions) we denote g by ∇f.

Definition 7.3. *Write $\mathcal{D}$ for the class of functions $f \in \mathcal{A} \cap L^2(T, \nu)$ such that $\nabla f \in L^2(T, \nu)$.*

Remark 7.4. By the observations made in Definition 7.2, the integral

$$\int_a^b \bar{g}(\xi|u) \, du$$

is well-defined for any $\xi \in E_+$ and $\bar{g} \in L^2(T, \nu)$.

Theorem 7.5. *The Dirichlet form $\mathcal{E}$ corresponding to the strongly continuous, self-adjoint, Markovian semigroup $(P_t)_{t \geqslant 0}$ on $L^2(T, \nu)$ has domain $\mathcal{D}$ and is given by*

$$\mathcal{E}(f,g) = \frac{1}{2} \int_T \nabla f(x) \nabla g(x) \, \nu(dx), \quad f,g \in \mathcal{D}. \qquad (7.3)$$

Proof. A virtual reprise of the argument in Example A.1 shows that the form $\mathcal{E}'$ given by the right–hand side of (7.3) is a Dirichlet form.

Let $(G_\alpha)_{\alpha>0}$ denote the resolvent corresponding to $(P_t)_{t\geq0}$: that is, $G_\alpha f = \int_0^\infty e^{-\alpha t} P_t f \, dt$ for $f \in L^2(T,\nu)$. In order to show that $\mathcal{E} = \mathcal{E}'$, it suffices to show that $G_\alpha(L^2(T,\nu)) \subseteq \mathcal{D}$ and $\mathcal{E}'_\alpha(G_\alpha f, g) := \mathcal{E}'(G_\alpha f, g) + \alpha(f, g) = (f, g)$ for $f \in L^2(T,\nu)$ and $g \in \mathcal{D}$, where we write $(\cdot, \cdot)$ for the $L^2(T,\nu)$ inner product. By a simple approximation argument, it further suffices to check that $G_\alpha(b\mathcal{B}(T) \cap L^2(T,\nu)) \subseteq \mathcal{D}$ and $\mathcal{E}'_\alpha(G_\alpha f, g) = (f, g)$ for $f \in b\mathcal{B}(T) \cap L^2(T,\nu)$ and $g \in \mathcal{D}$

Observe that

$$\int_0^\infty e^{-\alpha t} \phi_{a,t}(b, c) \, dt = 2 \exp\left(-\sqrt{2\alpha}(c - 2b + a) \right), \quad b < a \wedge c,$$

– see Equations 3.71.13 and 6.23.15 of [143]. Therefore, for $f \in b\mathcal{B}(T) \cap L^2(T,\nu)$ we have

$$G_\alpha f(x) = 2 \int_{-\infty}^{h(x)} \int_b^\infty \mu(x, b, c; f) e^{-\sqrt{2\alpha}(c - 2b + h(x))} \, dc \, db. \qquad (7.4)$$

Thus, $G_\alpha f \in \mathcal{A}$ with

$$\nabla(G_\alpha f)(x) = 2 \int_{h(x)}^\infty \mu(x, h(x), c; f) e^{-\sqrt{2\alpha}(c - h(x))} \, dc \qquad (7.5)$$
$$- \sqrt{2\alpha} G_\alpha f(x).$$

In order to show that $G_\alpha f \in \mathcal{D}$ is remains to show that the first term on the righ-hand side of (7.5) is in $L^2(T,\nu)$. By the Cauchy-Schwarz inequality and recalling the definition of T_t from (3.4),

$$\int_T \left[\int_{h(x)}^\infty \mu(x,h(x),c;f) e^{-\sqrt{2\alpha}(c-h(x))}\, dc \right]^2 \nu(dx)$$

$$= \int_{-\infty}^\infty \sum_{x \in T_a} \left[\int_a^\infty \frac{\int_{E_+} f(\xi|c)\mathbf{1}\{\xi|a = x\}\mu(d\xi)}{\mu\{\xi : \xi|a = x\}} e^{-\sqrt{2\alpha}(c-a)}\, dc \right]^2$$
$$\times \mu\{\xi : \xi|a = x\}\, da$$

$$\leqslant \frac{1}{2\sqrt{2\alpha}} \int_{-\infty}^\infty \sum_{x \in T_a} \left[\int_a^\infty \left[\frac{\int_{E_+} f(\xi|c)\mathbf{1}\{\xi|a = x\}\mu(d\xi)}{\mu\{\xi : \xi|a = x\}} \right]^2 e^{-\sqrt{2\alpha}(c-a)}\, dc \right]$$
$$\times \mu\{\xi : \xi|a = x\}\, da$$

$$\leqslant \frac{1}{2\sqrt{2\alpha}} \int_{-\infty}^\infty \sum_{x \in T_a} \left[\int_a^\infty \frac{\int_{E_+} f^2(\xi|c)\mathbf{1}\{\xi|a = x\}\mu(d\xi)}{\mu\{\xi : \xi|a = x\}} e^{-\sqrt{2\alpha}(c-a)}\, dc \right]$$
$$\times \mu\{\xi : \xi|a = x\}\, da$$

$$= \frac{1}{2\sqrt{2\alpha}} \int_{-\infty}^\infty \int_a^\infty \left[\int_{E_+} f^2(\xi|c)\mu(d\xi) \right] e^{-\sqrt{2\alpha}(c-a)}\, dc\, da$$

$$= \frac{1}{4\alpha} \int_{-\infty}^\infty \int_{E_+} f^2(\xi|c)\, dc\, \mu(d\xi) = \frac{1}{4\alpha} \int_T f^2(x)\, \nu(dx) < \infty,$$

as required.

From (7.5) we have for $g \in \mathcal{D}$ that

$$\mathcal{E}'(G_\alpha f, g)$$
$$= \int_{-\infty}^\infty \int_{E_+} \left[\int_a^\infty \mu(\xi|a, a, c; f) e^{-\sqrt{2\alpha}(c-a)}\, dc \right] \nabla g(\xi|a)\, \mu(d\xi)\, da \qquad (7.6)$$
$$- \frac{1}{2}\sqrt{2\alpha} \int_{-\infty}^\infty \int_{E_+} G_\alpha f(x) \nabla g(\xi|a), \mu(d\xi)\, da.$$

Consider the first term on the right–hand side of (7.6). Note that it can be written as

$$\int_{-\infty}^\infty \sum_{x \in T_a} \left[\int_a^\infty \frac{\int_{E_+} f(\xi|c)\mathbf{1}\{\xi|a = x\}\mu(d\xi)}{\mu\{\xi : \xi|a = x\}} e^{-\sqrt{2\alpha}(c-a)}\, dc \right]$$
$$\times \nabla g(x)\mu\{\xi : \xi|a = x\}\, da \qquad (7.7)$$
$$= \int_{-\infty}^\infty \int_{E_+} \left[\int_a^\infty f(x|c) e^{-\sqrt{2\alpha}(c-a)}\, dc \right] \nabla g(\xi|a)\, \mu(d\xi)\, da.$$

Substitute (7.7) into (7.6), integrate by parts, and use (7.5) to get that

$$\mathcal{E}'(G_\alpha f, g) = \int_{E_+} \int_{-\infty}^{\infty} f(\xi|a) g(\xi|a) \, da \, \mu(dx)$$

$$- \sqrt{2\alpha} \int_{E_+} \int_{-\infty}^{\infty} \left[\int_a^{\infty} f(x|c) e^{-\sqrt{2\alpha}(c-a)} \, dc \right] g(\xi|a) \, da \, \mu(d\xi)$$

$$+ \sqrt{2\alpha} \int_{E_+} \int_{-\infty}^{\infty} \left[\int_a^{\infty} \mu(\xi|a, a, c; f) e^{-\sqrt{2\alpha}(c-a)} \, dc \right] g(\xi|a) \, da \, \mu(d\xi)$$

$$- \alpha \int_{E_+} \int_{-\infty}^{\infty} G_\alpha f(\xi|a) g(\xi|a) \, da \, \mu(dx).$$

Argue as in (7.7) to see that the second and third terms on the right–hand side cancel and so

$$\mathcal{E}'(G_\alpha f, g) = (f, g) - \alpha(G_\alpha f, g),$$

as required.   □

Remark 7.6. We wish to apply to X the theory of symmetric processes and their associated Dirichlet forms developed in [72]. Because T is not generally locally compact, we need to to check that the conditions of Theorem A.8 hold – see Remark A.9.

We first show that conditions (a)–(c) of Theorem A.8 hold. That is, that there is a countably generated subalgebra $\mathcal{C} \subseteq bC(T) \cap \mathcal{D}$ such that $\mathcal{C}$ is $\mathcal{E}_1$-dense in $\mathcal{D}$, $\mathcal{C}$ separates points of T, and for each $x \in T$ there exists $f \in \mathcal{C}$ with $f(x) > 0$. Let $\mathcal{C}_0$ be a countable subset of $bC(T) \cap L^2(T, \nu)$ that separates points of T and is such that for every $x \in T$ there exists $f \in \mathcal{C}_0$ with $f(x) > 0$. Let $\mathcal{C}$ be the algebra generated by the countable collection $\bigcup_\alpha G_\alpha \mathcal{C}_0$, where the union is over the positive rationals. It is clear that $\mathcal{C}$ is $\mathcal{E}_1$-dense in $\mathcal{D}$. We observed in the proof of Theorem 7.1 that $P_t : bC(T) \to bC(T)$ for all $t \geq 0$ and $\lim_{t \downarrow 0} P_t f(x) = f(x)$ for all $f \in bC(T)$. Thus, $G_\alpha : bC(T) \to bC(T)$ for all $\alpha > 0$ and $\lim_{\alpha \to \infty} \alpha G_\alpha f(x) = f(x)$ for all $f \in bC(T)$. Therefore, $\mathcal{C}$ separates points of T and for every $x \in T$ there exists $f \in \mathcal{C}$ with $f(x) > 0$.

It remains to check that the tightness condition (d) of Theorem A.8 holds. That is, for all $\epsilon > 0$ there exists a compact set K such that $\mathrm{Cap}(T \backslash K) < \epsilon$ where Cap denotes the capacity associated with $\mathcal{E}_1$. However, it follows from the sample path continuity of X and Theorem IV.1.15 of [106] that, in the terminology of that result, the process X is ν-tight. Conditions IV.3.1 (i) – (iii) of [106] then hold by Theorem IV.5.1 of [106], and this suffices by Theorem III.2.11 of [106] to establish condition that (d) of Theorem A.8 holds.

7.4 Recurrence, Transience, and Regularity of Points

The Green operator G associated with the semigroup $(P_t)_{t \geq 0}$ is defined by $Gf(x) := \int_0^\infty P_t f(x) \, dt = \sup_{\alpha > 0} G_\alpha f(x)$ for $f \in p\mathcal{B}(T)$. In the terminology of [72], we say that X is *transient* is $Gf < \infty$, ν-a.e., for any $f \in L_+^1(T, \nu)$, whereas X is *recurrent* if $Gf \in \{0, \infty\}$, ν-a.e., for any $f \in L_+^1(T, \nu)$.

As we observed in Section 7.3, X has symmetric transition densities $p_t(x, y)$ with respect to ν such that $p_t(x, y) > 0$ for all $x, y \in T$. Consequently, in the terminology of [72], X is *irreducible*. Therefore, by Lemma 1.6.4 of [72], X is either transient or recurrent, and if X is recurrent, then $Gf = \infty$ for any $f \in L^1_+(T, \nu)$ that is not ν-a.e. 0.

Taking limits as $\alpha \downarrow 0$ in (7.4), we see that

$$Gf(x) = \int_T g(x, y) f(y) \, \nu(dy),$$

where

$$
\begin{aligned}
g(x, y) &:= 2 \int_{-\infty}^{h(x \wedge y)} \frac{1}{\mu\{\xi : \xi | b = x | b\}} \, db \\
&= 2 \int_{-\infty}^{h(x \wedge y)} \frac{1}{\mu\{\xi : \xi | b = y | b\}} \, db.
\end{aligned}
\tag{7.8}
$$

Note that the integrals

$$\int_{-\infty}^{a} \frac{1}{\mu\{\xi : \xi | b = \zeta | b\}} \, db, \ a \in \mathbb{R}, \ \zeta \in E_+,
\tag{7.9}$$

are either simultaneously finite or infinite. The following is now obvious.

Theorem 7.7. *If the integrals in (7.9) are finite (resp. infinite), then $g(x, y) < \infty$ (resp. $g(x, y) = \infty$) for all $x, y \in T$ and X is transient (resp. recurrent).*

Remark 7.8. For $B \in \mathcal{B}(T)$ write $\sigma_B := \inf\{t > 0 : X_t \in B\}$. We note from Theorem 4.6.6 and Problem 4.6.3 of [72] that if $\mathbb{P}^x\{\sigma_B < \infty\} > 0$ for some $x \in T$, then $\mathbb{P}^x\{\sigma_B < \infty\} > 0$ for all $x \in T$. Moreover, if X is recurrent, then $\mathbb{P}^x\{\sigma_B < \infty\} > 0$ for some $x \in T$ implies that $\mathbb{P}^x\{\forall N \in \mathbb{N}, \exists t > N : X_t \in B\} = 1$ for all $x \in T$.

Given $y \in T$, write σ_y for $\sigma_{\{y\}}$. Set $C = \{z \in T : y \leqslant z\}$. Pick $x \leqslant y$ with $x \neq y$. By definition of $(P_t)_{t \geqslant 0}$, $\mathbb{P}^x\{X_t \in C\} > 0$ for all $t > 0$. In particular, $\mathbb{P}^x\{\sigma_C < \infty\} > 0$. It follows from Axioms I and II that if $\gamma : \mathbb{R}_+ \mapsto T$ is any continuous map with $\{x, z\} \subset \gamma(\mathbb{R}_+)$ for some $z \in C$, then $y \in \gamma(\mathbb{R}_+)$ also. Therefore, by the sample path continuity of X, $\mathbb{P}^x\{\sigma_y < \infty\} > 0$ for this particular choice of x. However, Remark 7.8 then gives that $\mathbb{P}^x\{\sigma_y < \infty\} > 0$ for all $x \in T$. By Theorem 4.1.3 of [72] we have that points are regular for themselves. That is, $\mathbb{P}^x\{\sigma_x = 0\} = 1$ for all $x \in T$.

7.5 Examples

Recall the the family of $\mathbb{R}$–tree without ends (T, d) construction in Subsection 3.4.3 for a prime number p and constants $r_-, r_+ \geqslant 1$.

In the notation of Subsection 3.4.3, define a Borel measure μ on E_+ as follows. Write $\ldots \leqslant w_{-1} \leqslant w_0 = 1 \leqslant w_1 \leqslant w_2 \leqslant \ldots$ for the possible values of $w(\cdot,\cdot)$. That is, $w_k = \sum_{i=0}^{k} r_+^i$ if $k \geqslant 0$, whereas $w_k = 1 - \sum_{i=0}^{-k} r_-^i$ if $k < 0$. By construction, closed balls in E_+ all have diameters of the form 2^{-w_k} for some $k \in \mathbb{Z}$ and such a ball is the disjoint union of p balls of diameter $2^{-w_{k+1}}$. We can, therefore, uniquely define μ by requiring that each closed ball of diameter 2^{-w_k} has mass p^{-k}. The measure μ is nothing but the (unique up to constants) Haar measure on the locally compact Abelian group E_+.

Applying Theorem 7.7, we see that X will be transient if and only if $\sum_{k \in \mathbb{N}_0} p^{-k} r_-^k < \infty$, that is, if and only if $r_- < p$. As we might have expected, transience and recurrence are unaffected by the value of r_+: Theorem 7.7 shows that transience and recurrence are features of the structure of T "near" $\dagger$, whereas r_+ only dictates the structure of the T "near" points of E_+.

7.6 Triviality of the Tail σ–field

Theorem 7.9. *For all $x \in T$ the tail σ–field $\bigcap_{s \geqslant 0} \sigma\{X_t : t \geqslant s\}$ is $\mathbb{P}^x$–trivial (that is, consists of sets with $\mathbb{P}^x$–measure 0 or 1).*

Proof. Fix $x \in T$. By the continuity of the sample paths of X, $\sigma_{x|a} = \inf\{t > 0 : h(X_t) = a\}$. Because $h(X)$ is a Brownian motion, this stopping time is $\mathbb{P}^x$-a.s. finite. Put $T_0 := 0$ and $T_k := \sigma_{x|(h(x)-k)}$ for $k = 1, 2, \ldots$ By the strong Markov property we get that $\mathbb{P}^x\{T_1 < T_2 < \cdots < \infty\} = 1$. Set $X_k(t) := X((T_k + t) \wedge T_{k+1})$ for $k = 0, 1, \ldots$ Note that the tail σ-field in the statement of the result can also be written as $\bigcap_{k \geqslant 0} \sigma\{(T_\ell, X_\ell) : \ell \geqslant k\}$.

By the strong Markov property, the pairs $((T_{k+1} - T_k, X_k))_{k \in \mathbb{N}_0}$ are independent. Moreover, by the spatial homogeneity of Brownian motion, the random variables $(T_{k+1} - T_k)_{k \in \mathbb{N}_0}$ are identically distributed. The result now follows from Lemma 7.10 below. □

Lemma 7.10. *Let $\{(Y_n, Z_n)\}_{n \in \mathbb{N}}$ be a sequence of independent $\mathbb{R} \times \mathbf{U}$–valued random variables, where $(\mathbf{U}, \mathcal{U})$ is a measurable space. Suppose further that that the random variables Y_n, $n \in \mathbb{N}$, have a common distribution. Put $W_n := Y_1 + \ldots + Y_n$. Then the tail σ–field $\bigcap_{m \in \mathbb{N}} \sigma\{(W_n, Z_n) : n \geqslant m\}$ is trivial.*

Proof. Consider a real–valued random variable V that is measurable with respect to the tail σ–field in the statement. For each $m \in \mathbb{N}$ we have by conditioning on $\sigma\{W_n : n \geqslant m\}$ and using Kolmogorov's zero–one law that there is a $\sigma\{W_n : n \geqslant m\}$–measurable random variable V_m' such that $V_m' = V$ almost surely. Consequently, there is a random variable V' measurable with respect to $\bigcap_{m \in \mathbb{N}} \sigma\{W_n : n \geqslant m\}$ such that $V' = V$ almost surely, and the proof is completed by an application of the Hewitt–Savage zero–one law. □

Definition 7.11. *A function $f \in \mathcal{B}(T \times \mathbb{R}_+)$ (resp. $f \in \mathcal{B}(T)$) is said to be space–time harmonic (resp. harmonic) if $0 \leqslant f < \infty$ and $P_s f(\cdot, t) = f(\cdot, s+t)$ (resp. $P_s f = f$) for all $s, t \geqslant 0$.*

Remark 7.12. There does not seem to be a generally agreed upon convention for the use of the term "harmonic". It is often used for the analogous definition without the requirement that the function is non–negative, and $P_t f(x) = \mathbb{P}^x[f(X_t)]$ is sometimes replaced by $\mathbb{P}^x[f(X_\tau)]$ for suitable stopping times τ. Also, the terms *invariant* and *regular* are sometimes used.

The following is a standard consequence of the triviality of the tail σ–field and irreducibility of the process, but we include a proof for completeness.

Corollary 7.13. *There are no non–constant bounded space–time harmonic functions (and, a fortiori, no non–constant bounded harmonic functions).*

Proof. Suppose that f is a bounded space–time harmonic function. For each $x \in T$ and $s \geqslant 0$ the process $(f(X_t, s + t))_{t \geqslant 0}$ is a bounded $\mathbb{P}^x$–martingale. Therefore $\lim_{t \to \infty} f(X_t, s + t)$ exists $\mathbb{P}^x$-a.s. and $f(x, s) = \mathbb{P}^x[\lim_{t \to \infty} f(X_t, s + t)] = \lim_{t \to \infty} f(X_t, s + t)$, $\mathbb{P}^x$-a.s., by the triviality of the tail. By the Markov property and the fact that X has everywhere positive transition densities with respect to ν we get that $f(s, x) = f(t, y)$ for ν-a.e. y for each $t > s$, and it is clear from this that f is a constant. $\qquad\square$

Remark 7.14. The conclusion of Corollary 7.13 for harmonic functions has the following alternative probabilistic proof. By the arguments in the proof of Theorem 7.9 we have that if $n \in \mathbb{Z}$ is such that $n < h(x)$, then $\mathbb{P}^x\{\sigma_{x|n} < \sigma_{x|(n-1)} < \sigma_{x|(n-2)} < \cdots < \infty\} = 1$. Suppose that f is a bounded harmonic function. Then $f(x) = \mathbb{P}^x[\lim_{t \to \infty} f(X_t)] = \lim_{k \to \infty} f(x|(-k))$. Now note for each pair $x, y \in T$ that $x|(-k) = y|(-k)$ for $k \in \mathbb{N}$ sufficiently large.

7.7 Martin Compactification and Excessive Functions

Suppose in this section that X is transient. Recall that $f \in \mathcal{B}(T)$ is *excessive* for $(P_t)_{t \geqslant 0}$ if $0 \leqslant f < \infty$, $P_t f \leqslant f$, and $\lim_{t \downarrow 0} P_t f = f$ pointwise. Recall the definition of harmonic function from Section 7.6. In this section we will obtain an integral representation for the excessive and harmonic functions.

Fix $x_0 \in T$ and define $k : T \times T \to \mathbb{R}$, the corresponding *Martin kernel*, by

$$
k(x, y) := \frac{g(x, y)}{g(x_0, y)} = \frac{\int_{-\infty}^{h(x \wedge y)} \mu\{\xi : \xi|b = y|b\}^{-1}\, db}{\int_{-\infty}^{h(x_0 \wedge y)} \mu\{\xi : \xi|b = y|b\}^{-1}\, db}
$$
$$
= \frac{\int_{-\infty}^{h(x \wedge y)} \mu\{\xi : \xi|b = x|b\}^{-1}\, db}{\int_{-\infty}^{h(x_0 \wedge y)} \mu\{\xi : \xi|b = x_0|b\}^{-1}\, db}. \tag{7.10}
$$

Note that the function k is continuous in both arguments and

$$
0 < \mathbb{P}^x\{\sigma_{x_0} < \infty\} \leqslant k(x, y) = \frac{\mathbb{P}^x\{\sigma_y < \infty\}}{\mathbb{P}^{x_0}\{\sigma_y < \infty\}} \leqslant \mathbb{P}^{x_0}\{\sigma_x < \infty\}^{-1} < \infty.
$$

We can follow the standard approach to constructing a Martin compactification when there are well–behaved potential kernel densities (e.g. [94, 108]). That is, we choose a countable, dense subset $S \subset T$ and compactify T using the sort of Stone–Čech–like procedure described in Section 3.4.2 to obtain a metrizable compactification T^M such that a sequence $\{y_n\}_{n \in \mathbb{N}} \subset T$ converges if and only if $\lim_n k(x, y_n)$ exists for all $x \in T$. We discuss the analytic interpretation of the Martin compactification later in this section. We investigate the probabilistic features of the compactification and the connection with Doob h-transforms in Section 7.8. We first show that T^M coincides with the compactification $\overline{T}$ of Section 3.4.2.

Proposition 7.15. *The compact metric spaces $\overline{T}$ and T^M are homeomorphic, so that T^M can be identified with $T \cup E$. If we define*

$$k(x, \eta) := \frac{\int_{-\infty}^{h(x \wedge \eta)} \mu\{\xi : \xi|b = \eta|b\}^{-1}\, db}{\int_{-\infty}^{h(x_0 \wedge \eta)} \mu\{\xi : \xi|b = \eta|b\}^{-1}\, db}, \quad x \in T, \eta \in T \cup E_+,$$

and $k(x, \dagger) = 1$, then $k(x, \cdot)$ is continuous on $T \cup E$. Moreover,

$$\sup_{x \in B} \sup_{\eta \in T \cup E} k(x, \eta) < \infty$$

for all balls $B \subset T$.

Proof. The rest of the proof will be almost immediate once we show for a sequence $\{y_n\}_{n \in \mathbb{N}} \subset T$ that $\lim_n k(x, y_n)$ exists for all $x \in T$ if and only if $\lim_n h(x \wedge y_n)$ exists (in the extended sense) for all $x \in T$.

It is clear that if $\lim_n h(x \wedge y_n)$ exists for all $x \in T$, then $\lim_n k(x, y_n)$ exists for all $x \in T$.

Suppose, on the other hand, that $\lim_n k(x, y_n)$ exists for all $x \in T$ but $\lim_n h(x' \wedge y_n)$ does not exist for some $x' \in T$. Then we can find $\epsilon > 0$ and $a < h(x') - \epsilon$ such that $x'' := x'|a \in T$, $\liminf_n h(x' \wedge y_n) \leq a - \epsilon$, and $\limsup_n h(x' \wedge y_n) \geq a + \epsilon$. This implies that for any $N \in \mathbb{N}$ there exists $p, q \geq N$ such that $h(x'' \wedge y_p) = h(x' \wedge y_p)$ and $h(x'' \wedge y_q) = a < a + \epsilon/2 < h(x' \wedge y_q)$. Thus, we obtain the contradiction

$$\liminf_n \frac{k(x', y_n)}{k(x'', y_n)} = \liminf_n \frac{g(x', y_n)}{g(x'', y_n)} = 1,$$

while

$$\limsup_n \frac{k(x', y_n)}{k(x'', y_n)} = \limsup_n \frac{g(x', y_n)}{g(x'', y_n)}$$
$$\geq \frac{\int_{-\infty}^{a+\epsilon/2} \mu\{\xi : \xi|b = x'|b\}^{-1}\, db}{\int_{-\infty}^{a} \mu\{\xi : \xi|b = x'|b\}^{-1}\, db} > 1.$$

$\square$

The following theorem essentially follows from results in [108], with most of the work that is particular to our setting being the argument that the points of E_+ are, in the terminology of [108],. Unfortunately, the standing assumption in [108] is that the state–space is locally compact. The requirement for this hypothesis can be circumvented using the special features of our process, but checking this requires a fairly close reading of much of [108]. Later, more probabilistic or measure–theoretic, approaches to the Martin boundary such as [51, 74, 73, 86] do not require local compactness, but are rather less concrete and less pleasant to compute with. Therefore, we sketch the relevant arguments.

Definition 7.16. *An excessive function f is said to be a* potential *if*

$$\lim_{t \to \infty} P_t f = 0.$$

(The term purely excessive function *is also sometimes used.)*

Theorem 7.17. *If u is an excessive function, then there is a unique finite measure γ on $\overline{T} = T \cup E$ such that $u(x) = \int_{T \cup E} k(x, \eta)\,\gamma(d\eta)$, $x \in T$. Furthermore, u is harmonic (resp. a potential) if and only if $\gamma(T) = 0$ (resp. $\gamma(E) = 0$).*

Proof. From Theorem XII.17 in [43] there exists a sequence $\{f_n\}_{n \in \mathbb{N}}$ of bounded non–negative functions such that Gf_n is bounded for all n and $Gf_1(x) \leqslant Gf_2(x) \leqslant \ldots \leqslant Gf_n(x) \uparrow u(x)$ as $n \to \infty$ for all $x \in T$. Define a measure γ_n by $\gamma_n(dy) := g(x_0, y)f_n(y)\,\nu(dy)$, so that $Gf_n(x) = \int_T k(x, y)\,\gamma_n(dy)$. Note that $\gamma_n(T) = Gf_n(x_0) \leqslant u(x_0) < \infty$. We can think of $\{\gamma_n\}_{n \in \mathbb{N}}$ as a sequence of finite measures on the compact space $\overline{T}$ with bounded total mass. Therefore, there exists a subsequence $(n_\ell)_{\ell \in \mathbb{N}}$ such that $\gamma = \lim_\ell \gamma_{n_\ell}$ exists in the topology of weak convergence of finite measures on $\overline{T}$. By Proposition 7.15, each of the functions $k(x, \cdot)$ is bounded and continuous, and so

$$\int_{T \cup E} k(x, \eta)\,\gamma(d\eta) = \lim_\ell \int_{T \cup E} k(x, \eta)\,\gamma_{n_\ell}(d\eta)$$

$$= \lim_\ell \int_T k(x, y)\,\gamma_{n_\ell}(dy)$$

$$= \lim_\ell Gf_{n_\ell}(x) = u(x).$$

This completes the proof of existence. We next consider the the uniqueness claim.

Note first of all that the set of excessive functions is a cone; that is, it is closed under addition and multiplication by non-negative constants. This cone has an associated strong order: we say that $f \ll g$ for two excessive functions if $g = f + h$ for some excessive function h. As remarked in XII.34 of [43], for any two excessive functions f and g there is a greatest lower bound excessive function h such that $h \ll f$, $h \ll g$ and $h' \ll h$ for any other excessive function h' with $h' \ll f$ and $h' \ll g$. There is a similarly defined least upper bound. Moreover, if h and k are respectively the greatest lower bounds and least upper bounds of two excessive functions f and g, then $f + g = h + k$. Thus, the cone of excessive functions is a lattice in the strong order.

From Proposition 7.15, all excessive functions are bounded on balls and *a fortiori* ν–integrable on balls. Thus, the excessive functions are a subset of the separable, locally convex, topological vector space $L^1_{\text{loc}}(T, \nu)$ of locally ν-integrable functions equipped with the metrizable topology of $L^1(T, \nu)$ convergence on balls.

Consider the convex set of excessive functions u such that $u(x_0) = 1$. Any measure appearing in the representation of such a function u is necessarily a probability measure. Given a sequence $\{u_n\}_{n \in \mathbb{N}}$ of such functions, we can, by the weak compactness argument described above, find a subsequence $(u_{n_\ell})_{\ell \in \mathbb{N}}$ that converges bounded pointwise, and, therefore, also in $L^1_{\text{loc}}(T, \nu)$, to some limit u. Thus, the set of excessive functions u such that $u(x_0) = 1$ is convex, compact and metrizable.

An arbitrary excessive function is a non-negative multiple of an excessive function u with $u(x_0) = 1$. Consequently, the cone of excessive functions is a cone in a locally convex, separable, topological vector space with a compact and metrizable base and this cone is a lattice in the associated strong order. The Choquet uniqueness theorem – see Theorem X.64 of [43] – guarantees that every excessive function u with $u(x_0) = 1$ can be represented uniquely as an integral over the extreme points of the compact convex set of such functions.

Write k_η for the excessive function $k(\cdot, \eta)$, $\eta \in T \cup E$. The uniqueness claim will follow provided we can show for all $\eta \in T \cup E$ that the function k_η is an extreme point. That is, we must show that if $k_\eta = \int_{T \cup E} k_{\eta'}\, \gamma(d\eta')$ for some probability measure γ, then γ is necessarily the point mass at η.

Each of the functions k_y, $y \in T$, is clearly a potential. A direct calculation using (7.4), which we omit, shows that if $\xi \in E$, then $\alpha G_\alpha k_\xi = k_\xi$ for all $\alpha > 0$, and this implies that k_ξ is harmonic. Thus, $\lim_{t \to \infty} P_t k_\eta$ is either 0 or k_η depending on whether $\eta \in T$ or $\eta \in E$. In particular, if $k_\eta = \int_{T \cup E} k_{\eta'}\, \gamma(d\eta')$, then

$$\lim_{t \to \infty} P_t k_\eta = \int_{T \cup E} \lim_{t \to \infty} P_t k_{\eta'}\, \gamma(d\eta') = \int_E k_{\eta'}\, \gamma(d\eta').$$

Thus, γ must be concentrated on T if $\eta \in T$ and on E if $\eta \in E$.

Suppose now for $y \in T$ that

$$k_y(x) = \int_T k_{y'}(x)\, \gamma(dy')$$

or, equivalently, that

$$\frac{g(x, y)}{g(x_0, y)} = \int_T \frac{g(x, y')}{g(x_0, y')}\, \gamma(dy').$$

Thus, we have

$$\int_T g(x, y')\,\pi(dy') = \int_T g(x, y')\,\rho(dy')$$

where π is the measure $\delta_y/g(x_0, y)$ and ρ is the measure $\gamma/g(x_0, \cdot)$. Let g_α be the kernel corresponding to the operator G; that is,

$$Gf(x) = \int_T g_\alpha(x, y)f(y)\,\nu(dy).$$

It is straightforward to check that $\alpha G_\alpha G = G - G_\alpha$ (this is just a special instance of the resolvent equation). Thus

$$\int_T g_\alpha(x, y')\,\pi(dy') = \int_T g_\alpha(x, y')\,\rho(dy')$$

and

$$\int_T \int_T f(x)g_\alpha(x, y')\,\pi(dy')\,\nu(dx) = \int_T \int_T f(x)g_\alpha(x, y')\,\rho(dy')\,\nu(dx)$$

for any bounded continuous function f. Since g_α is symmetric,

$$\int_T f(x)g_\alpha(x, y')\,\nu(dx) = \int_T g_\alpha(y', x)f(x)\,\nu(dx).$$

Moreover,

$$\alpha \int_T g_\alpha(y', x)f(x)\,\nu(dx) \le \sup_{x \in T}|f(x)|$$

and

$$\lim_{\alpha \to \infty} \alpha \int_T g_\alpha(y', x)f(x)\,\nu(dx) = f(y')$$

for all $y' \in T$. Thus, $\int_T f(y')\,\pi(dy') = \int_T f(y')\,\rho(dy')$ for any bounded continuous function f, and $\pi = \rho$ as required. The argument we have just given is essentially a special case of the principle of masses – see, for example, Proposition 1.1 of [75].

Similarly, suppose for some $\xi \in E_+$ that $k_\xi(x) = \int_E k_{\xi'}(x)\,\gamma(d\xi)$. For $x \in T$ and $a > h(x \wedge \xi)$

$$k_\xi(x) \geq \mathbb{P}^x[k_\xi(X_{\sigma_{\xi|a}})]$$

$$= \frac{g(x, \xi|a)}{g(\xi|a, \xi|a)} k(\xi|a, \xi)$$

$$= \frac{\int_{-\infty}^{h(x \wedge (\xi|a))} \mu\{\zeta : \zeta|b = (\xi|a)|b\}^{-1} db}{\int_{-\infty}^{h(\xi|a)} \mu\{\zeta : \zeta|b = (\xi|a)|b\}^{-1} db}$$

$$\times \frac{\int_{-\infty}^{h((\xi|a) \wedge \xi)} \mu\{\zeta : \zeta = \xi|b\}^{-1} db}{\int_{-\infty}^{h(x_0 \wedge \xi)} \mu\{\zeta : \zeta|b = \xi|b\}^{-1} db}$$

$$= \frac{\int_{-\infty}^{h(x \wedge \xi)} \mu\{\zeta : \zeta|b = \xi|b\}^{-1} db}{\int_{-\infty}^{a} \mu\{\zeta : \zeta|b = \xi|b\}^{-1} db}$$

$$\times \frac{\int_{-\infty}^{a} \mu\{\zeta : \zeta = \xi|b\}^{-1} db}{\int_{-\infty}^{h(x_0 \wedge \xi)} \mu\{\zeta : \zeta|b = \xi|b\}^{-1} db}$$

$$= k_\xi(x).$$

Thus, $k_\xi(x) = \mathbb{P}^x[k_\xi(X_{\sigma_{\xi|a}})]$ for all a sufficiently large. On the other hand, a similar argument shows for $\xi' \in E_+ \backslash \{\xi\}$ that

$$k_{\xi'}(x) \geq \mathbb{P}^x[k_{\xi'}(X_{\sigma_{\xi|a}})]$$

and

$$\mathbb{P}^x[k_{\xi'}(X_{\sigma_{\xi|a}})] = \frac{\int_{-\infty}^{h(\xi \wedge \xi')} \mu\{\zeta : \zeta|b = \xi|b\}^{-1} db}{\int_{-\infty}^{a} \mu\{\zeta : \zeta|b = \xi|b\}^{-1} db} k_{\xi'}(x),$$

for sufficiently large a, where the right–hand side converges to 0 as $a \to 0$. Similarly, $\lim_{a \to \infty} \mathbb{P}^x[k_\dagger(X_{\sigma_{\xi|a}})] = 0$. This clearly shows that if $k_\xi = \int_E k_{\xi'} \gamma(d\xi')$, then γ cannot assign any mass to $E \backslash \{\xi\}$. Uniqueness for the representation of $k_\dagger$ is handled similarly and the proof of the uniqueness claim is complete.

Lastly, the claim regarding representation of harmonic functions and potentials is immediate from what we have already shown.

$\square$

Remark 7.18. Theorem 7.17 can be used as follows to give an analytic proof (in the transient case) of the conclusion of Corollary 7.13 that bounded harmonic functions are necessarily constant.

First extend the definition of the Green kernel g to $T \cup E$ by setting

$$g(\eta, \rho) := 2 \int_{-\infty}^{h(\eta \wedge \rho)} \mu\{\zeta : \zeta|b = \eta|b\}^{-1} db$$

$$= 2 \int_{-\infty}^{h(\eta \wedge \rho)} \mu\{\zeta : \zeta|b = \rho|b\}^{-1} db.$$

By Theorem 7.17, non–constant bounded harmonic functions exist if and only if there is a non–trivial finite measure γ concentrated on E_+ such that

$$\sup_{x \in T} \int_{E_+} k(x, \zeta)\, \gamma(d\zeta) < \infty. \tag{7.11}$$

Note that for any ball $B \subset E_+$ of the form $B = \{\zeta \in E_+ : \zeta | h(x^*) = x^*\}$ for $h(x^*) \geqslant h(x_0)$ we have $g(x_0, \zeta) = g(x_0, x^*)$. Thus, by possibly replacing the measure γ in (7.11) by its trace on a ball, we have that non–constant bounded harmonic functions exist if and only if there is a probability measure (that we also denote by γ) concentrated on a ball $B \subset E_+$ such that

$$\sup_{x \in T} \int_B g(x, \zeta)\, \gamma(d\zeta) < \infty. \tag{7.12}$$

Observe that $g(\xi | t, \zeta)$ increases monotonically to $g(\xi, \zeta)$ as $t \to \infty$ and so, by monotone convergence, (7.12) holds if and only if

$$\sup_{\xi \in E_+} \int_B g(\xi, \zeta)\, \gamma(d\zeta) < \infty. \tag{7.13}$$

It is further clear that if (7.13) holds, then

$$\int_B \int_B g(\xi, \zeta)\, \gamma(d\xi)\, \gamma(d\zeta) < \infty. \tag{7.14}$$

Suppose that (7.14) holds. For $b \in \mathbb{R}$ write T_b^γ for the subset of T_b consisting of $x \in T_b$ such that $\gamma\{\xi \in B : \eta | b = x\} > 0$. In other words, T_b^γ is the collection of points of the form $\eta | b$ for some η in the closed support of γ. Note that $\sum_{x \in T_b^\gamma} \mu\{\eta : \eta | b = x\} \leqslant \mu(B)$ if 2^{-b} is at most the diameter of B. Applying Jensen's inequality, we obtain the contradiction

$$\int_B \int_B g(\xi, \zeta)\, \gamma(d\xi)\, \gamma(d\zeta)$$

$$= 2 \int_{-\infty}^\infty \int_B \int_B \frac{1\{\xi | b = \zeta | b\}}{\mu\{\eta : \eta | b = \xi | b\}}\, \gamma(d\xi)\, \gamma(d\zeta)\, db$$

$$= 2 \int_{-\infty}^\infty \int_B \frac{\gamma\{\eta : \eta | b = \xi | b\}}{\mu\{\eta : \eta | b = \xi | b\}}\, \gamma(d\xi)\, db$$

$$\geqslant 2 \int_{-\infty}^\infty \left[\int_B \frac{\mu\{\eta : \eta | b = \xi | b\}}{\gamma\{\eta : \eta | b = \xi | b\}}\, \gamma(d\xi) \right]^{-1} db$$

$$= 2 \int_{-\infty}^\infty \left[\sum_{x \in T_b^\gamma} \frac{\mu\{\eta : \eta | b = x\}}{\gamma\{\eta : \eta | b = x\}}\, \gamma\{\eta : \eta | b = x\} \right]^{-1} db$$

$$= \infty.$$

7.8 Probabilistic Interpretation of the Martin Compactification

Suppose that X is transient and consider the harmonic functions $k_\xi = k(\cdot, \xi)$, $\xi \in E_+$, introduced in Section 7.7 and the corresponding *Doob h-transform* laws $\mathbb{P}^x_{k_\xi}$, $x \in T$. That is, $\mathbb{P}^x_{k_\xi}$, $x \in T$, is the collection of laws of a Markov process X^ξ such that $\mathbb{P}^x_{k_\xi}[f(X^\xi_t)] = k_\xi(x)^{-1}\mathbb{P}^x[k_\xi(X_t)f(X_t)]$, $f \in b\mathcal{B}(T)$. The following result says that the process X^ξ can be thought of as "X conditioned to converge to ξ."

Theorem 7.19. *For all $x \in T$, $\mathbb{P}^x_{k_\xi}\{\lim_{t\to\infty} X^\xi_t = \xi\} = 1$.*

Proof. Note that X^ξ has Green kernel $k_\xi(x)^{-1}g(x,y)k_\xi(y) < \infty$. Thus, X^ξ is transient.

Now observe that $\lim_{t\to\infty} X^\xi_t$ exists. This is so because, by compactness, the limit exists along a subsequence and if two subsequences had different limits then there would be a ball in T that was visited infinitely often – contradicting transience.

Thus, it suffices to show that if $a > h(x \wedge \xi)$, then $\mathbb{P}^x_{k_\xi}\{\sigma_{\xi|a} < \infty\} = 1$. However, after some algebra,

$$\mathbb{P}^x_{k_\xi}\{\sigma_{\xi|a} < \infty\} = k_\xi(x)^{-1}\mathbb{P}^x[k_\xi(X_{\xi|a})\mathbf{1}\{\sigma_{\xi|a} < \infty\}]$$
$$= \frac{1}{k(x,\xi)} \frac{g(x,\xi|a)}{g(\xi|a,\xi|a)}k(\xi|a,\xi) = 1.$$

□

Remark 7.20. Recall that $(h(X_t))_{t\geq 0}$ is a standard Brownian motion under $\mathbb{P}^x$. We can ask what $(h(X^\xi_t))_{t\geq 0}$ looks like under $\mathbb{P}^x_{k_\xi}$. Arguing as in the proof of Theorem 7.24 below and using Girsanov's theorem, we have under $\mathbb{P}^x_{k_\xi}$ that

$$h(X^\xi_t) = h(X^\xi_0) + W_t + D_t,$$

where W is a standard Brownian motion and

$$D_t = \int_0^t \left[\frac{\mathbf{1}\{X_s \leqslant \xi\}}{\mu\{\zeta : X_s \leqslant \zeta\}} \right] \Big/ \left[\int_{-\infty}^{h(X_s)} \frac{1}{\mu\{\zeta : X_s|b \leqslant \zeta\}} \, db \right] ds.$$

In other words, when X^ξ_t is not on the ray R_ξ the height process $h(X^\xi_t)$ evolves as a standard Brownian motion, but when X^ξ_t is on the ray $R_\xi := \{x \in T : x \leqslant \xi\}$ the height experiences an added positive drift toward ξ.

7.9 Entrance Laws

A *probability entrance law* for the semigroup $(P_t)_{t\geq 0}$ is a family $(\gamma_t)_{t>0}$ of probability measures on T such that $\gamma_s P_t = \gamma_{s+t}$ for all $s, t > 0$. Given such a probability entrance law, we can construct on some probability space $(\Omega, \mathcal{F}, \mathbb{P})$ a continuous process that, with a slight abuse of notation, we denote $X = (X_t)_{t>0}$ such that X_t has law γ_t and X is a time–homogeneous Markov process with transition semigroup $(P_t)_{t\geq 0}$.

In this section we show that the only probability entrance laws are the trivial ones (that is, there is no way to start the process "from infinity" in some sense).

Theorem 7.21. *If $(\gamma_t)_{t>0}$ is a probability entrance law for $(P_t)_{t\geqslant 0}$, then $\gamma_t = \gamma_0 P_t$, $t > 0$, for some probability measure γ_0 on T.*

Proof. Construct a Ray–Knight compactification (T^R, ρ), say, as in Section 17 of [128]. Write $(\bar{P}_t)_{t\geqslant 0}$ and $(\bar{G}_\alpha)_{\alpha>0}$ for the corresponding extended semigroup and resolvent.

Construct X with one–dimensional distributions $(\gamma_t)_{t>0}$ and semigroup $(P_t)_{t\geqslant 0}$ as described above. By Theorem 40.4 of [128], $\lim_{t\downarrow 0} X_t$ exists in the Ray topology, and if γ_0 denotes the law of this limit, then $\gamma_0 \bar{P}_t$ is concentrated on T for all $t > 0$ and γ_t is the restriction of $\gamma_0 \bar{P}_t$ to T. We need, therefore, to establish that γ_0 is concentrated on T. Moreover, it suffices to consider the case when γ_0 is a point mass at some $x_0 \in T^R$, so that $\lim_{t\downarrow 0} X_t = x_0$ in the Ray topology. Note by Theorem 4.10 of [128] that the germ σ-field $\mathcal{F}_{0+} := \bigcap_\epsilon \sigma\{X_t : 0 \leqslant t \leqslant \epsilon\}$ is trivial under $\mathbb{P}$ in this case.

By construction of $(P_t)_{t\geqslant 0}$, the family obtained by pushing forward each γ_t by the map h is an entrance law for standard Brownian motion on $\mathbb{R}$. Because Brownian motion is a Feller–Dynkin process, the only entrance laws for it are the trivial ones $(\rho Q_t)_{t>0}$, where $(Q_t)_{t\geqslant 0}$ is the semigroup of Brownian motion and ρ is a probability measure on $\mathbb{R}$. Thus, by the triviality $\mathcal{F}_{0+}$, there is a constant $h_0 \in \mathbb{R}$ such that $\lim_{t\downarrow 0} h(X_t) = h_0$, $\mathbb{P}$-a.s.

As usual, regard functions on T as functions on T^R by extending them to be 0 on $T^R \backslash T$. For every $f \in b\mathcal{B}(T)$ we have by Theorem 40.4 of [128] that $\lim_{t\downarrow 0} G_\alpha f(X_t) = \lim_{t\downarrow 0} \bar{G}_\alpha f(X_t) = \bar{G}_\alpha f(x)$.

From (7.4),

$$G_\alpha f(x) = \int_T g_\alpha(x, y) f(y) \, \nu(dy),$$

where

$$\begin{aligned} g_\alpha(x,y) &:= 2 \int_{-\infty}^{h(x \wedge y)} \frac{\exp(-\sqrt{2\alpha}(h(x) + h(y) - 2b))}{\mu\{\xi : \xi|b = x|b\}} \, db \\ &= 2 \int_{-\infty}^{h(x \wedge y)} \frac{\exp(-\sqrt{2\alpha}(h(x) + h(y) - 2b))}{\mu\{\xi : \xi|b = y|b\}} \, db. \end{aligned} \tag{7.15}$$

It follows straightforwardly that $\lim_{t\downarrow 0} h(X_t \wedge y)$ exists for all $y \in T$, $\mathbb{P}$-a.s., and so, by the discussion in Section 3.4.2 and the triviality of $\mathcal{F}_{0+}$, there exists $\eta \in T \cup E$ such that $h(\eta) \leqslant h_0$ and $\lim_{t\downarrow 0} h(X_t \wedge y) = h(\eta \wedge y)$, $\mathbb{P}$-a.s. Note, in particular, that we actually have $\eta \in T \cup \{\dagger\}$ because $h(\eta) < \infty$. Moreover, we conclude that

$$\begin{aligned} \int_0^\infty e^{-\alpha t} \gamma_t(f) \, dt &= \bar{G}_\alpha f(x_0) \\ &= 2 \int_T \left[\int_{-\infty}^{h(\eta \wedge y)} \frac{\exp(-\sqrt{2\alpha}(h_0 + h(y) - 2b))}{\mu\{\xi : \xi|b = y|b\}} \, db \right] \nu(dy) \end{aligned}$$

for all $f \in b\mathcal{B}(T)$.

We cannot have $\eta = \dagger$, because this would imply that γ_t is the null measure for all $t > 0$. If $\eta \in T$ and $h_0 = h(\eta)$, then we have $\gamma_t = \delta_\eta P_t$.

We need, therefore, only rule out the possibility that $\eta \in T$ but $h(\eta) < h_0$. In this case we have

$$\int_0^\infty e^{-\alpha t} \gamma_t(f)\, dt = \exp\left(-\sqrt{2\alpha}(h_0 - h(\eta))\right) \int_0^\infty e^{-\alpha t} \delta_\eta P_t(f)\, dt$$

and so, by comparison of Laplace transforms, $\gamma_t = \int_0^t \delta_\eta P_{t-s}\, \kappa(ds)$, where κ is a certain stable-$\frac{1}{2}$ distribution. In particular, γ_t has total mass $\kappa([0,t]) < 1$ and is not a probability distribution. $\square$

7.10 Local Times and Semimartingale Decompositions

Our aim in this section is to give a semimartingale decomposition for the process $H_\xi(t) := h(X_t \wedge \xi)$, $t \geq 0$, for $\xi \in E_+$.

This result will be analogous to the classical Tanaka's formula for a standard Brownian motion B that says

$$B(t)_+ = B(0)_+ + \int_0^t \mathbf{1}\{B(s) > 0\}\, dB(s) + \frac{1}{2}\ell(t),$$

where ℓ is the *local time* of the Brownian motion at 0. In other words, B_+ is constant (at 0) over time intervals when $B < 0$ and during time intervals when $B \geq 0$ it evolves like a standard Brownian motion except at 0 when it gets an additive positive "kick" from the local time.

From the intuitive description of X in the Section 7.1, we similarly expect H_ξ to remain constant over time intervals when X_t is not in the ray $R_\xi := \{x \in T : x \leq \xi\}$. During time intervals when X_t is in R_ξ we expect H_ξ to evolve as a standard Brownian motion except at branch points of T where it receives negative "kicks" from a local time additive functional. Here the magnitude of the kicks will be related to how much μ–mass is being lost to the rays that are branching off from R_ξ.

To make this description precise, we first need to introduce the appropriate local time processes and then use Fukushima's stochastic calculus for Dirichlet processes (in much the same way that Tanaka's formula follows from the standard Itô's formula for Brownian motion). Unfortunately, this involves appealing to quite a large body of material from [72], but it would have required lengthening this section considerably to state in detail the results that we use.

We showed in Section 7.4 that $\mathbb{P}^x\{\sigma_y < \infty\}$ for any $x, y \in T$. By Theorems 4.2.1 and 2.2.3 of [72], the point mass δ_y at any $y \in T$ belongs to the set of measures S_{00}. (See (2.2.10) of [72] for a definition of S_{00}. Another way of seeing that δ_y is in S_{00} is just to observe that $\sup_x g_\alpha(x, y) < \infty$ for all $\alpha > 0$.)

By Theorem 5.1.6 of [72] there exists for each $y \in T$ a strict sense positive continuous additive functional L^y with Revuz measure δ_y. As usual, we call L^y the *local time* at y.

Definition 7.22. *Given $\xi \in E_+$, write m_ξ for the Radon measure on T that is supported on the ray R_ξ and for each $a \in \mathbb{R}$ assigns mass $\mu\{\zeta \in E_+ : \zeta|a = \xi|a\}$ to the set $\{\xi|b : b \geqslant a\} = \{x \in R_\xi : h(x) \geqslant a\}$.*

Remark 7.23. Note that m_ξ is a discrete measure that is concentrated on the countable set of points of the form $\xi \wedge \zeta$ for some $\zeta \in E_+ \backslash \{\xi\}$ (that is, on the points where other rays branch from R_ξ).

Theorem 7.24. *For each $\xi \in E_+$ and $x \in T$ the process H_ξ has a semimartingale decomposition*

$$H_\xi(t) = H_\xi(0) + M_\xi(t) - \frac{1}{2} \int_{R_\xi} L^y(t)\, m_\xi(dy), \ t \geqslant 0,$$

under $\mathbb{P}^x$, where M_ξ is a continuous, square–integrable martingale with quadratic variation

$$\langle M_\xi \rangle(t) = \int_0^t \mathbf{1}\{X(s) \leqslant \xi\}\, ds, \ t \geqslant 0.$$

Moreover, the martingales M_ξ and $M_{\xi'}$ for $\xi, \xi' \in E_+$ have covariation

$$\langle M_\xi, M_{\xi'} \rangle_t = \int_0^t \mathbf{1}\{X(s) \leqslant \xi \wedge \xi'\}\, ds, \ t \geqslant 0.$$

Proof. For $\xi \in E_+$, $x \in T$, and $A \in \mathbb{N}$, set $h_\xi(x) = h(x \wedge \xi)$ and $h_\xi^A(x) = (-A) \vee (h(x \wedge \xi) \wedge A)$.

It is clear that h_ξ^A is in the domain $\mathcal{D}$ of the Dirichlet form $\mathcal{E}$, with $\nabla h_\xi^A(x) = \mathbf{1}\{\xi|(-A) \leqslant x \leqslant \xi|A\}$. Given $f \in \mathcal{D}$, it follows from the product rule that

$$2\mathcal{E}(h_\xi^A f, h_\xi^A f) - \mathcal{E}((h_\xi^A)^2, f) = \int_T f(x)\mathbf{1}\{\xi|(-A) \leqslant x \leqslant \xi|A\}\, \nu(dx).$$

In the terminology of Section 3.2 of [72], the *energy measure* corresponding to h_ξ^A is $\nu_\xi^A(dx) := \mathbf{1}\{\xi|(-A) \leqslant x \leqslant \xi|A\}\, \nu(dx)$. A similar calculation shows that the joint energy measure corresponding to a pair of functions h_ξ^A and $h_{\xi'}^{A'}$ is

$$\mathbf{1}[\{\xi|(-A) \leqslant x \leqslant \xi|A\} \cap \{\xi'|(-A') \leqslant x \leqslant \xi'|A'\}]\, \nu(dx) = (\nu_\xi^A \wedge \nu_{\xi'}^{A'})(dx)$$ in the usual lattice structure on measures.

An integration by parts establishes that for any $f \in \mathcal{D}$ we have

$$\mathcal{E}(h_\xi^A, f) = \frac{1}{2} \int_T f(x)\, \tilde{m}_\xi^A(dx),$$

where

$$\tilde{m}^A_\xi := m^A_\xi - \mu\{\zeta : \zeta|(-A) = \xi|(-A)\}\delta_{\xi|(-A)} + \mu\{\zeta : \zeta|A = \xi|A\}\delta_{\xi|A}$$

with

$$m^A_\xi(dx) := \mathbf{1}\{\xi|(-A) \leqslant x \leqslant \xi|A\}m_\xi(dx).$$

Now ν^A_ξ is the Revuz measure of the strict sense positive continuous additive functional $\int_0^t \mathbf{1}\{\xi|(-A) \leqslant X(s) \leqslant \xi|A\}\,ds$ and $\nu^A_\xi \wedge \nu^{A'}_{\xi'}$ is the Revuz measure of the strict sense positive continuous additive functional $\int_0^t \mathbf{1}[\{\xi|(-A) \leqslant X(s) \leqslant \xi|A\} \cap \{\xi'|(-A') \leqslant X(s) \leqslant \xi'|A'\}]\,ds$. A straightforward calculation shows that $\sup_x \int g_\alpha(x, y)\,m^A_\xi(dy) < \infty$, and so $m^A_\xi \in S_{00}$ is the Revuz measure of the strict sense positive continuous additive functional $\int_{R_\xi} L^y(t)\,m^A_\xi(dy)$ (because the integral is just a sum, we do not need to address the measurability of $y \mapsto L^y(t)$).

Put $H^A_\xi(t) := h^A_\xi(X(t))$, $t \geqslant 0$. Theorem 5.2.5 of [72] applies to give that

$$H^A_\xi(t) = H^A_\xi(0) + M^A_\xi(t) - \frac{1}{2}\int_{R_\xi} L^y(t)\,\tilde{m}^A_\xi(dy), \; t \geqslant 0,$$

under $\mathbb{P}^x$ for each $x \in T$, where M^A_ξ is a continuous, square–integrable martingale with quadratic variation

$$\langle M^A_\xi \rangle(t) = \int_0^t \mathbf{1}\{\xi|(-A) \leqslant X(s) \leqslant \xi|A\}\,ds.$$

Moreover, the martingales M^A_ξ and $M^{A'}_{\xi'}$ for $\xi, \xi' \in E_+$ have covariation

$$\langle M^A_\xi, M^{A'}_{\xi'} \rangle(t)$$
$$= \int_0^t \mathbf{1}\left[\{\xi|(-A) \leqslant X(s) \leqslant \xi|A\} \cap \{\xi'|(-A') \leqslant X(s) \leqslant \xi'|A'\}\right]\,ds.$$

In particular,

$$\langle M^B_\xi - M^A_\xi \rangle(t)$$
$$= \int_0^t \mathbf{1}\left[\{\xi|(-B) \leqslant X(s) \leqslant \xi|B\}\backslash\{\xi|(-A) \leqslant X(s) \leqslant \xi|A\}\right]\,ds \qquad (7.16)$$

for $A < B$.

For each $t \geqslant 0$ we have that $H^A_\xi(s) = H_\xi(s)$ and $\int_{R_\xi} L^y(s)\,\tilde{m}^A_\xi(dy) = \int_{R_\xi} L^y(s)\,m_\xi(dy)$ for all $0 \leqslant s \leqslant t$ when $A > \sup\{|H_\xi(s)| : 0 \leqslant s \leqslant t\}$, $\mathbb{P}^x$-a.s. Therefore, there exists a continuous process M_ξ such that $M^A_\xi(s) = M_\xi(s)$ for all $0 \leqslant s \leqslant t$ when $A > \sup\{|H_\xi(s)| : 0 \leqslant s \leqslant t\}$, $\mathbb{P}^x$-a.s. It follows from (7.16) that $\lim_{A\to\infty} \mathbb{P}^x[\sup_{0\leqslant s\leqslant t}|M^A_\xi(s) - M_\xi(s)|^2] = 0$. By standard arguments, the processes M_ξ are continuous, square–integrable martingales with the stated quadratic variation and covariation properties. $\qquad\square$

Remark 7.25. There is more that can be said about the process H_ξ. For instance, given $x \in T$ and $\xi \in E_+$ with $x \in R_\xi$ and $a > h(x)$, we can explicitly calculate the Laplace transform of $\inf\{t > 0 : H_\xi(t) = a\} = \sigma_{\xi|a}$ under $\mathbb{P}^x$. We have

$$\mathbb{P}^x[\exp(-\alpha\sigma_{\xi|a})] = g_\alpha(x, \xi|a) \,/\, g_\alpha(\xi|a, \xi|a),$$

where g_α is given explicitly by (7.15). When X is transient, the distribution of $\sigma_{\xi|a}$ has an atom at ∞ and we have

$$\mathbb{P}^x \left\{ \sup_{0 \leqslant t < \infty} H_\xi(t) \geqslant a \right\} = \mathbb{P}^x\{\sigma_{\xi|a} < \infty\} = g(x, \xi|a) \,/\, g(\xi|a, \xi|a).$$

By the strong Markov property, the càdlàg process $(\sigma_{\xi|a})_{a \geqslant h(x)}$ has independent (although, of course, non–stationary) increments under $\mathbb{P}^x$, with the usual appropriate definition of this notion for non–decreasing $\mathbb{R} \cup \{+\infty\}$–valued processes.

ℝ–Trees from Coalescing Particle Systems

8.1 Kingman's Coalescent

Here is a quick description of *Kingman's coalescent* (which we will hereafter simply refer to as the coalescent). Let $\mathcal{P}$ denote the collection of partitions of ℕ. For $n \in \mathbb{N}$ let $\mathcal{P}_n$ denote the collection of partitions of $\mathbb{N}_{\leqslant n} := \{1, 2, \dots, n\}$. Write ρ_n for the natural restriction map from $\mathcal{P}$ onto $\mathcal{P}_n$. Kingman [90] showed that there was a (unique in law) $\mathcal{P}$–valued Markov process Π such that for all $n \in \mathbb{N}$ the restricted process $\Pi_n := \rho_n \circ \Pi$ is a $\mathcal{P}_n$–valued, time–homogeneous Markov chain with initial state $\Pi_n(0)$ the trivial partition $\{\{1\}, \dots, \{n\}\}$ and the following transition rates: if Π_n is in a state with k blocks, then

- a jump occurs at rate $\binom{k}{2}$,
- the new state is one of the $\binom{k}{2}$ partitions that can be obtained by merging two blocks of the current state,
- and all such possibilities are equally likely.

Let $N(t)$ denote the number of blocks of the partition $\Pi(t)$. It was shown in [90] that almost surely, $N(t) < \infty$ for all $t > 0$ and the process N is a pure–death Markov chain that jumps from k to $k - 1$ at rate $\binom{k}{2}$ for $k > 1$ (the state 1 is a trap). Therefore, the construction in Example 3.41 applies to construct a compact ℝ-tree from Π. Let $(\mathbb{S}, \delta)$ denote the corresponding (random) ultrametric space that arises from looking at the closure of the leaves (that is, ℕ) in that tree, as in Example 3.41. We note that some properties of the space $(\mathbb{N}, \delta)$ were considered explicitly in Section 4 of [10]. We will apply Proposition B.3 to show that the Hausdorff and packing dimensions of $\mathbb{S}$ are both 1 and that, in the terminology of [112] – see, also, [27, 113, 114] – the space $\mathbb{S}$ is a.s. *capacity–equivalent* to the unit interval $[0, 1]$.

Theorem 8.1. *Almost surely, the Hausdorff and packing dimensions of the random compact metric space $\mathbb{S}$ are both 1. There exist random variables C^*, C^{**} such that almost surely $0 < C^* \leqslant C^{**} < \infty$ and for every gauge f*

$$C^* \mathrm{Cap}_f([0,1]) \leqslant \mathrm{Cap}_f(\mathbb{S}) \leqslant C^{**} \mathrm{Cap}_f([0,1]).$$

Proof. We will apply Proposition B.3.

Note that $\sigma_n := \inf\{t > 0 : N(t) = n\}$ is of the form $\tau_{n+1} + \tau_{n+2} + \cdots$, where the τ_k are independent and τ_k is exponential with rate $\binom{k}{2}$. Thus

$$\mathbb{P}[\sigma_n] = \frac{2}{(n+1)n} + \frac{2}{(n+2)(n+1)} + \cdots = \frac{2}{n}. \qquad (8.1)$$

It is easy to check that

$$\lim_{t\downarrow 0} t N(t) = \lim_{n\to\infty} \sigma_n N(\sigma_n) = \lim_{n\to\infty} \sigma_n n = 2, \ a.s.$$

– see, for example, the arguments that lead to Equation (35) in [18].

It was shown in [90] that almost surely for all $t > 0$ the *asymptotic block frequencies*

$$F_i(t) := \lim_{n\to\infty} n^{-1} \big| \{ j \in \mathbb{N}_{\leqslant n} : j \sim_{\Pi(t)} I_i(t) \} \big|, \ 1 \leqslant i \leqslant N(t),$$

exist and

$$F_1(t) + \cdots + F_{N(t)}(t) = 1.$$

We claim that

$$\lim_{t\downarrow 0} t^{-1} \sum_{i=1}^{N(t)} F_i(t)^2 = 1, \ a.s. \qquad (8.2)$$

To see this, set $X_{n,i} := F_i(\sigma_n)$ for $n \in \mathbb{N}$ and $1 \leqslant i \leqslant n$, and observe from (8.1) that it suffices to establish

$$\lim_{n\to\infty} n \sum_{i=1}^{n} X_{n,i}^2 = 2, \ a.s. \qquad (8.3)$$

By the "paintbox" construction in Section 5 of [90] the random variable $\sum_{i=1}^{n} X_{n,i}^2$ has the same law as $U_{(1)}^2 + (U_{(2)} - U_{(1)})^2 + \cdots + (U_{(n-1)} - U_{(n-2)})^2 + (1 - U_{(n-1)})^2$, where $U_{(1)} \leqslant \ldots \leqslant U_{(n-1)}$ are the order statistics corresponding to i.i.d. random variables $U_1, \ldots, U_{n-1}$ that are uniformly distributed on $[0,1]$ – see Figure 8.1 and Section 4.2 of [18] for an exposition from which essentially this figure was taken with permission. By a classical result on the spacings between order statistics of i.i.d. uniform random variables – see, for example, Section III.3.(e) of [66] – the law of $\sum_{i=1}^{n} X_{n,i}^2$ is the same as that of $(\sum_{i=1}^{n} T_i^2)/(\sum_{i=1}^{n} T_i)^2$, where $T_1, \ldots, T_n$ are i.i.d. mean one exponential random variables.

Now for any $0 < \varepsilon < 1$ we have, recalling $\mathbb{P}[T_i^2] = 2$,

$$\mathbb{P}\left\{ \left(\sum_{i=1}^{n} T_i^2\right) \Big/ \left(\sum_{i=1}^{n} T_i\right)^2 > (1+\varepsilon)(1-\varepsilon)^{-2} 2n^{-1} \right\}$$

$$\leqslant \mathbb{P}\left\{ \sum_{i=1}^{n} (T_i^2 - \mathbb{P}[T_i^2]) > 2\varepsilon n \right\} + \mathbb{P}\left\{ \sum_{i=1}^{n} (T_i - \mathbb{P}[T_i]) < -\varepsilon n \right\}.$$

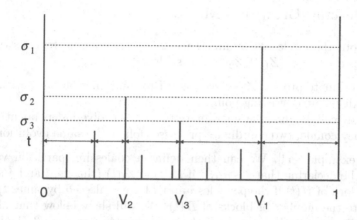

Fig. 8.1. Kingman's description of the block frequencies in the coalescent. Let $V_1, V_2, \ldots$ be independent random variables uniformly distributed on $[0, 1]$. For $\sigma_n \leqslant t < \sigma_{n-1}$ put $Y_1(t) = V_{(1)}, Y_2(t) = V_{(2)} - V_{(1)}, \ldots, Y_n(t) = 1 - V_{(n-1)}$, where $V_{(1)}, \ldots, V_{(n-1)}$ are the order statistics of $V_1, \ldots V_{n-1}$. Then, as set valued processes, the block proportions $(\{F_1(t), \ldots F_{N(t)}(t)\})_{t \geqslant 0}$ and the spacings $(\{Y_1(t), \ldots Y_{N(t)}(t)\})_{t \geqslant 0}$ have the same distribution.

A fourth moment computation and Markov's inequality show that both terms on the right–hand side are bounded above by $c(\varepsilon)n^{-2}$ for a suitable constant $c(\epsilon)$. A similar bound holds for

$$\mathbb{P}\left\{ \left(\sum_{i=1}^{n} T_i^2 \right) \Big/ \left(\sum_{i=1}^{n} T_i \right)^2 < (1-\varepsilon)(1+\varepsilon)^{-2} 2n^{-1} \right\}.$$

The claim (8.3) and, hence, (8.2) now follows by an application of the Borel–Cantelli Lemma.

The proof is finished by an appeal to Proposition B.3 and the observation there exist constants $0 < c^{\#} \leqslant c^{\#\#} < \infty$ such that

$$c^{\#} \left(\int_0^1 f(t)\, dt \right)^{-1} \leqslant \mathrm{Cap}_f([0,1]) \leqslant c^{\#\#} \left(\int_0^1 f(t)\, dt \right)^{-1}$$

(this is described as "classical" in [113] and follows by arguments similar to those used in Section 3 of that paper to prove a higher dimensional analogue of this fact).   □

8.2 Coalescing Brownian Motions

Let $\mathbb{T}$ denote the circle of circumference 2π. It is possible to construct a stochastic process $\mathbf{Z} = (Z_1(t), Z_2(t), \ldots)$ such that:

- each coordinate process Z_i evolves as a Brownian motion on $\mathbb{T}$ with uniformly distributed starting point,
- until they collide, different coordinate processes evolve independently,
- after they collide, two coordinate processes follow the same evolution

– see, for example, [44]. We can then define a coalescing partition valued process Π be declaring that $i \sim_{\Pi(t)} j$ if $Z_i(t) = Z_j(t)$ (that is, i and j are in the same block of $\Pi(t)$ if the particles i and j have coalesced by times t). Let $N(t)$ denote the number of blocks of $\Pi(t)$. We will show below that almost surely $N(t) < \infty$ for all $t > 0$, and the procedure in Example 3.41 gives a ℝ-tree with leaves corresponding to $\mathbb{N}$ and a compactification of $\mathbb{N}$ that we will denote by $(\mathbb{S}, \delta)$.

Our main result is the following.

Theorem 8.2. *Amost surely, the random compact metric space $(\mathbb{S}, \delta)$ has Hausdorff and packing dimensions both equal to $\frac{1}{2}$. There exist random variables K^*, K^{**} such that $0 < K^* \leqslant K^{**} < \infty$ and for every gauge f*

$$K^* \mathrm{Cap}_f(C_{\frac{1}{2}}) \leqslant \mathrm{Cap}_f(\mathbb{S}) \leqslant K^{**} \mathrm{Cap}_f(C_{\frac{1}{2}}),$$

where $C_{\frac{1}{2}}$ is the middle-$\frac{1}{2}$ Cantor set.

Remark 8.3. One of the assertions of the following result is that $\mathbb{S}$ is a.s. capacity–equivalent to $C_{\frac{1}{2}}$. Hence, by the results of [113], $\mathbb{S}$ is also a.s. capacity–equivalent to the zero set of (one–dimensional) Brownian motion.

Before proving Theorem 8.2, we will need to do some preliminary computations to enable us to check the conditions of Proposition B.3.

Given a finite non–empty set $A \subseteq \mathbb{T}$, let W^A be a process taking values in the space of finite subsets of $\mathbb{T}$ that describes the evolution of a finite set of indistinguishable Brownian particles with the features that $W^A(0) = A$ and that particles evolve independently between collisions but when two particles collide they coalesce into a single particle.

Write $\mathcal{O}$ for the collection of open subsets of $\mathbb{T}$ that are either empty or consist of a finite union of open intervals with distinct end–points. Given $B \in \mathcal{O}$, define on some probability space $(\Sigma, \mathcal{G}, \mathbb{Q})$ an $\mathcal{O}$–valued process V^B, the *annihilating circular Brownian motion* as follows. The end–points of the constituent intervals execute independent Brownian motions on $\mathbb{T}$ until they collide, at which point they annihilate each other. If the two colliding end–points are from different intervals, then those two intervals merge into one interval. If the two colliding end–points are from the same interval, then that interval vanishes (unless the interval was arbitrarily close to $\mathbb{T}$ just before the collision, in which case the process takes the value $\mathbb{T}$). The process is stopped when it hits the empty set or $\mathbb{T}$.

We have the following duality relation between W^A and V^B. An analogous result for the coalescing Brownian flow on $\mathbb{R}$ is on p18 of [22].

Proposition 8.4. *For all finite, non–empty subsets $A \subseteq \mathbb{T}$, all sets $B \in \mathcal{O}$, and all $t \geqslant 0$,*

$$\mathbb{P}\{W^A(t) \subseteq B\} = \mathbb{Q}\{A \subseteq V^B(t)\}.$$

Proof. For $N \in \mathbb{N}$, let $\mathbb{Z}_N := \{0, 1, \ldots N - 1\}$ denote the integers modulo N. Let $\mathbb{Z}_N^{\frac{1}{2}} := \{\frac{1}{2}, \frac{3}{2}, \ldots, \frac{2N-1}{2}\}$ denote the half–integers modulo N. A non–empty subset D of $\mathbb{Z}_N$ can be (uniquely) decomposed into "intervals": an interval of D is an equivalence class for the equivalence relation on the points of D defined by $x \sim y$ if and only if $x = y$, $\{x, x+1, \ldots, y-1, y\} \subseteq D$, or $\{y, y+1, \ldots, x-1, x\} \subseteq D$ (with all arithmetic modulo N). Any interval other than $\mathbb{Z}_N$ itself has an associated pair of (distinct) "end–points" in $\mathbb{Z}_N^{\frac{1}{2}}$: if the interval is $\{a, a+1, \ldots, b-1, b\}$, then the corresponding end–points are $a - \frac{1}{2}$ and $b + \frac{1}{2}$ (with all arithmetic modulo N). Note that the end–points of different intervals of D are distinct.

For $C \subseteq \mathbb{Z}_N$, let W_N^C be a process on some probability space $(\Omega', \mathcal{F}', \mathbb{P}')$ taking values in the collection of non–empty subsets of $\mathbb{Z}_N$ that is defined in the same manner as W^A, with Brownian motion on $\mathbb{T}$ replaced by simple, symmetric (continuous time) random walk on $\mathbb{Z}_N$ (that is, by the continuous time Markov chain on $\mathbb{Z}_N$ that only makes jumps from x to $x + 1$ or x to $x - 1$ at a common rate $\lambda > 0$ for all $x \in \mathbb{Z}_N$). For $D \subseteq \mathbb{Z}_N$, let V_N^D be a process taking values in the collection of subsets of $\mathbb{Z}_N$ that is defined on some probability space $(\Sigma', \mathcal{G}', \mathbb{Q}')$ in the same manner as V^B, with Brownian motion on $\mathbb{T}$ replaced by simple, symmetric (continuous time) random walk on $\mathbb{Z}_N^{\frac{1}{2}}$ (with the same jump rate λ as in the definition of W_N^C). That is, end–points of intervals evolve as annihilating random walks on $\mathbb{Z}_N^{\frac{1}{2}}$.

The proposition will follow by a straightforward weak limit argument if we can show the following duality relationship between the coalescing "circular" random walk W_N^C and the annihilating "circular" random walk V_N^D:

$$\mathbb{P}'\{W_N^C(t) \subseteq D\} = \mathbb{Q}'\{C \subseteq V_N^D(t)\} \tag{8.4}$$

for all non–empty subsets of $C \subseteq \mathbb{Z}_N$, all subsets of $D \subseteq \mathbb{Z}_N$, and all $t \geqslant 0$.

It is simple, but somewhat tedious, to establish (8.4) by a generator calculation using the usual generator criterion for duality – see, for example, Corollary 4.4.13 of [56]. However, as Tom Liggett pointed out to us, there is an easier route. A little thought shows that V_N^D is nothing other than the (simple, symmetric) voter model on $\mathbb{Z}_N$. The analogous relationship between the annihilating random walk and the voter model on $\mathbb{Z}$ due to [124] is usually called the *border equation* – see Section 2 of [32] for a discussion and further references. The relationship (8.4) is then just the analogue of the usual duality between the voter model and coalescing random walk on $\mathbb{Z}$ and it can be established in a similar manner by Harris's graphical method (again see Section 2 of [32] for a discussion and references and Figure 8.2 for an illustration).

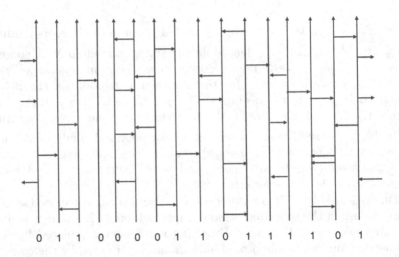

$$0 \quad 1 \quad 1 \quad 0 \quad 0 \quad 0 \quad 0 \quad 1 \quad 1 \quad 0 \quad 1 \quad 1 \quad 1 \quad 1 \quad 0 \quad 1$$

Fig. 8.2. The graphical construction of the (symmetric, nearest neighbor) voter model on $\mathbb{Z}_{16}$. Time proceeds up the page. The initial configuration is at the bottom of the diagram. Horizontal arrows issue from each site at rate λ, and are equally likely to point left or right. The state of the site at the head of an arrow is changed to the current state of the site at the tail. Arrows wrap around modulo 16. Going forwards in time, the boundaries between blocks of 0s and blocks of 1s execute a family of continuous time annihilating simple random walks. By reversing the direction of the vertical and horizontal arrows, it is possible to trace back from some location in space and time to the ultimate origin at time 0 of the state at that location. The resulting history is a continuous time simple random walk. Any two such histories evolve independently until they collide, after which they coalesce.

□

Define a set–valued processes $W^{[n]}$, $n \in \mathbb{N}$, and W by

$$W^{[n]}(t) := \{Z_1(t), Z_2(t), \ldots, Z_n(t)\} \subseteq \mathbb{T}, \ t \geqslant 0,$$

and

$$W(t) := \{Z_1(t), Z_2(t), \ldots\} \subseteq \mathbb{T}, \ t \geqslant 0.$$

Thus, $W^{[1]}(t) \subseteq W^{[2]}(t) \subseteq \cdots$, $\bigcup_{n \in \mathbb{N}} W^{[n]}(t) = W(t)$, and the cardinality of $W(t)$ is $N(t)$, the number of blocks in the partition $\Pi(t)$.

Corollary 8.5. *For $t > 0$,*

$$\mathbb{P}[N(t)] = 1 + 2 \sum_{n \in \mathbb{N}} \exp\left(-\left(\frac{n}{2}\right)^2 t\right) < \infty$$

and

$$\lim_{t \downarrow 0} t^{\frac{1}{2}} \mathbb{P}[N(t)] = 2\sqrt{\pi}.$$

Proof. Note that if B is a single open interval (so that for all $t \geqslant 0$ the set $V^B(t)$ is either an interval or empty) and we let $L(t)$ denote the length of $V^B(t)$, then L is a Brownian motion on $[0, 2\pi]$ with infinitesimal variance 2 that is stopped at the first time it hits $\{0, 2\pi\}$.

Now, for $M \in \mathbb{N}$ and $0 \leqslant i \leqslant M - 1$ we have from the translation invariance of Z and Proposition 8.4 that

$$\mathbb{P}\left\{W^{[n]}(t) \cap [2\pi i/M, 2\pi(i+1)/M] \neq \varnothing\right\}$$
$$= 1 - \mathbb{P}\left\{W^{[n]}(t) \subseteq]0, 2\pi(M-1)/M[\right\}$$
$$= 1 - \mathbb{P}\left\{W^{[n]}(0) \subseteq V^{]0, 2\pi(M-1)/M[}(t)\right\},$$

where we take the annihilating process $V^{]0, 2\pi(M-1)/M[}$ to be defined on the same probability space $(\Omega, \mathcal{F}, \mathbb{P})$ as the process $\mathbf{Z}$ that was used to construct $W^{[n]}$ and W, and we further take the processes $V^{]0, 2\pi(M-1)/M[}$ and $\mathbf{Z}$ to be independent. Thus,

$$\mathbb{P}\{W(t) \cap [2\pi i/M, 2\pi(i+1)/M] \neq \varnothing\}$$
$$= 1 - \mathbb{P}\left\{V^{]0, 2\pi(M-1)/M[}(t) = \mathbb{T}\right\}$$
$$= 1 - \tilde{\mathbb{P}}\left\{\tilde{\tau} \leqslant 2t, \ \tilde{B}(\tilde{\tau}) = 2\pi \mid \tilde{B}(0) = 2\pi(M-1)/M\right\},$$

where $\tilde{B}$ is a standard one–dimensional Brownian motion on some probability space $(\tilde{\Omega}, \tilde{\mathcal{F}}, \tilde{\mathbb{P}})$ and $\tilde{\tau} = \inf\{s \geqslant 0 : \tilde{B}(s) \in \{0, 2\pi\}\}$.

By Theorem 4.1.1 of [91] we have

$$\mathbb{P}\left[\,|W(t)|\,\right]$$

$$= \lim_{M \to \infty} \mathbb{P}\left[\sum_{i=0}^{M-1} \mathbf{1}\left\{W(t) \cap [2\pi i/M, 2\pi(i+1)/M] \neq \varnothing\right\}\right]$$

$$= \lim_{M \to \infty} M\left(1 - \tilde{\mathbb{P}}\left\{\tilde{\tau} \leqslant 2t,\ \tilde{B}(\tilde{\tau}) = 2\pi \mid \tilde{B}(0) = 2\pi(M-1)/M\right\}\right)$$

$$= 1 - \lim_{M \to \infty} M\frac{2}{\pi}\sum_{n \in \mathbb{N}} \frac{(-1)^n}{n} \sin\left(n\pi\left(\frac{M-1}{M}\right)\right) \exp\left(-\left(\frac{n}{2}\right)^2 t\right)$$

$$= 1 + 2\sum_{n \in \mathbb{N}} \exp\left(-\left(\frac{n}{2}\right)^2 t\right)$$

$$= \theta\left(\frac{t}{4\pi}\right) < \infty,$$

where

$$\theta(u) := \sum_{n=-\infty}^{\infty} \exp(-\pi n^2 u) \tag{8.5}$$

is the Jacobi theta function (we refer the reader to [31] for a survey of many of the other probabilistic interpretations of the theta function). The proof is completed by recalling that θ satisfies the functional equation $\theta(u) = u^{-\frac{1}{2}}\theta(u^{-1})$ and noting that $\lim_{u \to \infty} \theta(u) = 1$. □

For $t > 0$ the random partition $\Pi(t)$ is exchangeable with a finite number of blocks. Let $1 = I_1(t) < I_2(t) < \cdots < I_{N(t)}(t)$ be the list in increasing order of the minimal elements of the blocks of $\Pi(t)$. Results of Kingman – see Section 11 of [11] for a unified account – and the fact that Π evolves by pairwise coalescence of blocks give that $\mathbb{P}$–a.s. for all $t > 0$ the asymptotic frequencies

$$F_i(t) = \lim_{n \to \infty} n^{-1}|\{j \in \mathbb{N}_{\leqslant n} : j \sim_{\Pi(t)} I_i(t)\}|$$

exist for $1 \leqslant i \leqslant N(t)$ and $F_1(t) + \cdots + F_{N(t)}(t) = 1$.

Lemma 8.6. *Almost surely,*

$$\lim_{t \downarrow 0} t^{-\frac{1}{2}} \sum_{i=1}^{N(t)} F_i(t)^2 = \frac{2}{\pi^{3/2}}.$$

Proof. Put $T_{ij} := \inf\{t \geqslant 0 : Z_i(t) = Z_j(t)\}$ for $i \neq j$. Observe that

$$\mathbb{P}\left[\sum_{i=1}^{N(t)} F_i(t)^2\right] = \mathbb{P}\left[\lim_{n \to \infty} \frac{1}{n^2} \sum_{i=1}^{n} \sum_{k=1}^{n} \mathbf{1}\left\{j \sim_{\Pi(t)} k\right\}\right]$$

$$= \mathbb{P}\{1 \sim_{\Pi(t)} 2\}$$

$$= \mathbb{P}\{T_{12} \leqslant t\}.$$

From Theorem 4.1.1 of [91] we have

$$\mathbb{P}\{T_{12} \leqslant t\}$$

$$= \frac{1}{2\pi} \int_0^{2\pi} 1 - \frac{4}{\pi} \sum_{n \in \mathbb{N}} \sin\left(\frac{(2n-1)x}{2}\right) \frac{1}{2n-1} \exp\left(-\left(\frac{2n-1}{2}\right)^2 t\right) dx$$

$$= \frac{8}{\pi^2} \sum_{n \in \mathbb{N}} \frac{1}{(2n-1)^2} \left\{1 - \exp\left(-\left(\frac{2n-1}{2}\right)^2 t\right)\right\}$$

$$= \frac{2}{\pi^2} \int_0^t \sum_{n \in \mathbb{N}} \exp\left(-\left(\frac{2n-1}{2}\right)^2 s\right) ds$$

$$= \frac{2}{\pi^2} \int_0^t \frac{1}{2} \left\{\sum_{n=-\infty}^{\infty} \exp\left(-n^2 \frac{s}{4}\right) - \sum_{n=-\infty}^{\infty} \exp\left(-n^2 s\right)\right\} ds$$

$$= \frac{1}{\pi^2} \int_0^t \left\{\theta\left(\frac{s}{4\pi}\right) - \theta\left(\frac{s}{\pi}\right)\right\} ds,$$

where θ is again the Jacobi theta function defined in (8.5). By the properties of θ recalled after (8.5),

$$\lim_{t \downarrow 0} t^{-\frac{1}{2}} \mathbb{P}\left[\sum_{i=1}^{N(t)} F_i(t)^2\right] = \lim_{t \downarrow 0} t^{-\frac{1}{2}} \mathbb{P}\{T_{12} \leqslant t\} = \frac{2}{\pi^{3/2}}. \tag{8.6}$$

Now

$$\mathbb{P}\left[\left(\sum_{i=1}^{N(t)} F_i(t)^2\right)^2\right]$$

$$= \mathbb{P}\left[\lim_{n \to \infty} \frac{1}{n^4} \sum_{i_1=1}^{n} \sum_{i_2=1}^{n} \sum_{i_3=1}^{n} \sum_{i_4=1}^{n} \mathbf{1}\left\{i_1 \sim_{\Pi(t)} i_2, i_3 \sim_{\Pi(t)} i_4\right\}\right]$$

$$= \mathbb{P}\{1 \sim_{\Pi(t)} 2, 3 \sim_{\Pi(t)} 4\},$$

and so

$$\mathrm{Var}\left(\sum_{i=1}^{N(t)} F_i(t)^2\right) = \mathbb{P}\{1 \sim_{\Pi(t)} 2, 3 \sim_{\Pi(t)} 4\} - \mathbb{P}\{T_{12} \leqslant t\}^2$$

$$= \mathbb{P}\{1 \sim_{\Pi(t)} 2, 3 \sim_{\Pi(t)} 4\} - \mathbb{P}\{T_{12} \leqslant t, T_{23} \leqslant t\}. \tag{8.7}$$

Observe that

$$\mathbb{P}\{T_{12} \leqslant t, T_{34} \leqslant t, T_{13} > t, T_{14} > t, T_{23} > t, T_{24} > t\}$$

$$\leqslant \mathbb{P}\{1 \sim_{\Pi(t)} 2, 3 \sim_{\Pi(t)} 4, |W^{[4]}(t)| \neq 1\}$$

$$\leqslant \mathbb{P}\{T_{12} \leqslant t, T_{34} \leqslant t\}$$

and

$$\mathbb{P}\{T_{12} \leqslant t, T_{34} \leqslant t\} - \mathbb{P}\{T_{12} \leqslant t, T_{34} \leqslant t, T_{13} > t, T_{14} > t, T_{23} > t, T_{24} > t\}$$
$$\leqslant \sum_{i=1,2} \sum_{j=3,4} \mathbb{P}\{T_{12} \leqslant t, T_{34} \leqslant t, T_{ij} \leqslant t\}.$$

Thus

$$\mathrm{Var}\left(\sum_{i=1}^{N(t)} F_i(t)^2\right) \leqslant \mathbb{P}\{1 \sim_{\Pi(t)} 2 \sim_{\Pi(t)} 3 \sim_{\Pi(t)} 4\}$$

$$+ \sum_{i=1,2} \sum_{j=3,4} \mathbb{P}\{T_{12} \leqslant t, T_{34} \leqslant t, T_{ij} \leqslant t\}. \tag{8.8}$$

Put $D_{ij} := |Z_i(0) - Z_j(0)|$. We have

$$\mathbb{P}\{1 \sim_{\Pi(t)} 2 \sim_{\Pi(t)} 3 \sim_{\Pi(t)} 4\}$$
$$= \mathbb{P}\{T_{12} \leqslant t, T_{13} \wedge T_{23} \leqslant t, T_{14} \wedge T_{24} \wedge T_{34} \leqslant t\}$$
$$= \mathbb{P}\Bigg(\{T_{12} \leqslant t, T_{13} \wedge T_{23} \leqslant t, T_{14} \wedge T_{24} \wedge T_{34} \leqslant t\}$$

$$\setminus \{D_{12} \leqslant t^{\frac{2}{5}}, (D_{13} \wedge D_{23}) \leqslant t^{\frac{2}{5}}, (D_{14} \wedge D_{24} \wedge D_{34}) \leqslant t^{\frac{2}{5}}\}\Bigg) \tag{8.9}$$

$$+ \mathbb{P}\{D_{12} \leqslant t^{\frac{2}{5}}, (D_{13} \wedge D_{23}) \leqslant t^{\frac{2}{5}}, (D_{14} \wedge D_{24} \wedge D_{34}) \leqslant t^{\frac{2}{5}}\}$$

$$\leqslant \sum_{1 \leqslant i < j \leqslant 4} \mathbb{P}\{T_{ij} \leqslant t, D_{ij} > t^{\frac{2}{5}}\} + \mathbb{P}\left\{\max_{1 \leqslant i < j \leqslant 4} D_{ij} \leqslant 3t^{\frac{2}{5}}\right\},$$

where we have appealed to the triangle inequality in the last step. Because $\frac{2}{5} < \frac{1}{2}$, an application of the reflection principle and Brownian scaling certainly gives that the probability $\mathbb{P}\{T_{ij} \leqslant t, D_{ij} > t^{\frac{2}{5}}\}$ is $o(t^{\alpha})$ as $t \downarrow 0$ for any $\alpha > 0$. Moreover, by the translation invariance of m (the common distribution of the $Z_i(0)$), the second term in the rightmost member of (8.9) is at most

$$\mathbb{P}\{|Z_2(0) - Z_1(0)| \leqslant 3t^{\frac{2}{5}}, |Z_3(0) - Z_1(0)| \leqslant 3t^{\frac{2}{5}}, |Z_4(0) - Z_1(0)| \leqslant 3t^{\frac{2}{5}}\}$$
$$= \mathbb{P}\{|Z_2(0)| \leqslant 3t^{\frac{2}{5}}, |Z_3(0)| \leqslant 3t^{\frac{2}{5}}, |Z_4(0)| \leqslant 3t^{\frac{2}{5}}\}$$
$$= ct^{\frac{6}{5}},$$

for a suitable constant c when t is sufficiently small. Therefore,

$$\mathbb{P}\{1 \sim_{\Pi(t)} 2 \sim_{\Pi(t)} 3 \sim_{\Pi(t)} 4\}$$
$$= \mathbb{P}\{\{T_{12} \leqslant t, T_{13} \wedge T_{23} \leqslant t, T_{14} \wedge T_{24} \wedge T_{34} \leqslant t\} \tag{8.10}$$
$$= O(t^{\frac{6}{5}}), \quad \text{as } t \downarrow 0.$$

A similar argument establishes that

$$\mathbb{P}\{T_{12} \leqslant t, T_{34} \leqslant t, T_{ij} \leqslant t\} = O(t^{\frac{6}{5}}), \quad \text{as } t \downarrow 0, \tag{8.11}$$

for $i = 1, 2$ and $j = 3, 4$.

Substituting (8.10) and (8.11) into (8.8) gives

$$\text{Var}\left(\sum_{i=1}^{N(t)} F_i(t)^2\right) = O(t^{\frac{6}{5}}), \quad \text{as } t \downarrow 0.$$

This establishes the desired result when combined with the expectation calculation (8.6), Chebyshev's inequality, a standard Borel–Cantelli argument, and the monotonicity of $\sum_{i=1}^{N(t)} F_i(t)^2$. $\square$

We may suppose that on our probability space $(\Omega, \mathcal{F}, \mathbb{P})$ there is a sequence $B_1, B_2, \ldots$ of i.i.d. one–dimensional standard Brownian motions with initial distribution the uniform distribution on $[0, 2\pi]$ and that Z_i is defined by setting $Z_i(t)$ to be the image of $B_i(t)$ under the usual homomorphism from $\mathbb{R}$ onto $\mathbb{T}$. For $n \in \mathbb{N}$ and $0 \leqslant j \leqslant 2^n - 1$, let $I_1^{n,j} \leqslant I_2^{n,j} \leqslant \ldots$ be a list in increasing order of the set of indices $\{i \in \mathbb{N} : B_i(0) \in [2\pi j/2^n, 2\pi(j+1)/2^n[\}$. Put $B_i^{n,j} := B_{I_i^{n,j}}$ and $Z_i^{n,j} := Z_{I_i^{n,j}}$. Thus, $(B_i^{n,j})_{i\in\mathbb{N}}$ is an i.i.d. sequence of standard $\mathbb{R}$–valued Brownian motions and $(Z_i^{n,j})_{i\in\mathbb{N}}$ is an i.i.d. sequence of standard $\mathbb{T}$–valued Brownian motions. In each case the corresponding initial distribution is uniform on $[2\pi j/2^n, 2\pi(j+1)/2^n[$. Moreover, for $n \in \mathbb{N}$ fixed the sequences $(B_i^{n,j})_{i\in\mathbb{N}}$ are independent as j varies and the same is true of the sequences $(Z_i^{n,j})_{i\in\mathbb{N}}$.

Let $\underline{W}$ (resp. $\underline{W}^{n,j}$, $W^{n,j}$) be the coalescing system defined in terms of $(B_i)_{i\in\mathbb{N}}$ (resp. $(B_i^{n,j})_{i\in\mathbb{N}}$, $(Z_i^{n,j})_{i\in\mathbb{N}}$) in the same manner that W is defined in terms of $(Z_i)_{i\in\mathbb{N}}$.

It is clear by construction that

$$N(t) = |W(t)| \leqslant \sum_{i=0}^{2^n-1} |W^{n,i}(t)| \leqslant \sum_{i=0}^{2^n-1} |\underline{W}^{n,i}(t)|, \quad t > 0, n \in \mathbb{N}. \tag{8.12}$$

Lemma 8.7. *The expectation $\mathbb{P}[|\underline{W}(1)|]$ is finite.*

Proof. There is an obvious analogue of the duality relation Proposition 8.4 for systems of coalescing and annihilating one–dimensional Brownian motions. Using this duality and arguing as in the proof of Corollary 8.5, it is easy to see that, letting $\bar{L}$ and $\bar{U}$ be two independent, standard, real-valued Brownian motions on some probability space $(\bar{\Omega}, \bar{\mathcal{F}}, \bar{\mathbb{P}})$ with $\bar{L}(0) = \bar{U}(0) = 0$,

$\mathbb{P}[|\underline{W}(1)|]$

$$= \lim_{M \to \infty} \sum_{i=-\infty}^{\infty} \mathbb{P}\{\underline{W}(1) \cap [2\pi i/M, 2\pi(i+1)/M] \neq \varnothing\}$$

$$= \lim_{M \to \infty} \sum_{i=-\infty}^{\infty} \bar{\mathbb{P}}\Big\{ \min_{0 \leqslant t \leqslant 1} \big((\bar{U}(t) + 2\pi(i+1)/M) - (\bar{L}(t) + 2\pi i/M)\big) > 0,$$

$$[\bar{L}(1) + 2\pi i/M, \bar{U}(1) + 2\pi(i+1)/M] \cap [0, 2\pi] \neq \varnothing\Big\}$$

$$\leqslant \limsup_{M \to \infty} c'M\bar{\mathbb{P}}\left[\mathbf{1}\Big\{ \min_{0 \leqslant t \leqslant 1} (\bar{U}(t) - \bar{L}(t)) > -2\pi/M \Big\} (\bar{U}(1) - \bar{L}(1) + c'')\right]$$

for suitable constants c' and c''. Noting that $(\bar{U} - \bar{L})/\sqrt{2}$ is a standard Brownian motion, the result follows from a straightforward calculation with the joint distribution of the minimum up to time 1 and value at time 1 of such a process – see, for example, Corollary 30 in Section 1.3 of [70].   □

Proposition 8.8. *Almost surely,*

$$0 < \liminf_{t \downarrow 0} t^{\frac{1}{2}} N(t) \leqslant \limsup_{t \downarrow 0} t^{\frac{1}{2}} N(t) < \infty.$$

Proof. By the Cauchy–Schwarz inequality,

$$1 = \left(\sum_{i=1}^{N(t)} F_i(t)\right)^2 \leqslant N(t) \sum_{i=1}^{N(t)} F_i(t)^2.$$

Hence, by Lemma 8.6,

$$\liminf_{t \downarrow} t^{\frac{1}{2}} N(t) \geqslant \frac{\pi^{\frac{3}{2}}}{2}, \quad \mathbb{P} - a.s.$$

On the other hand, for each $n \in \mathbb{N}$, $|\underline{W}^{n,i}(2^{-2n})|$, $i = 0, \ldots, 2^n - 1$, are i.i.d. random variables that, by Brownian scaling, have the same distribution as $|\underline{W}(1)|$. By (8.12),

$$t^{\frac{1}{2}} N(t) \leqslant \frac{1}{2^{n-1}} \sum_{i=0}^{2^n - 1} |\underline{W}^{n,i}(2^{-2n})|$$

for $2^{-2n} < t \leqslant 2^{-2(n-1)}$. An application of Lemma 8.7 and the following strong law of large numbers for triangular arrays completes the proof.   □

Lemma 8.9. *Consider a triangular array* $\{X_{n,i} : 1 \leqslant i \leqslant 2^n, n \in \mathbb{N}\}$ *of identically distributed, real–valued, mean zero, random variables on some probability space* $(\Omega, \mathcal{F}, \mathbb{P})$ *such that the collection* $\{X_{n,i} : 1 \leqslant i \leqslant 2^n\}$ *is independent for each* $n \in \mathbb{N}$. *Then*

$$\lim_{n \to \infty} 2^{-n}(X_{n,1} + \cdots + X_{n,2^n}) = 0, \quad \mathbb{P} - a.s.$$

Proof. This sort of result appears to be known in the theory of complete convergence. For example, it follows from the much more general Theorem A in [23] by taking $N_n = 2^n$ and $\psi(t) = 2^t$ in the notation of that result – see also the Example following that result. For the sake of completeness, we give a short proof that was pointed out to us by Michael Klass.

Let $\{Y_n : n \in \mathbb{N}\}$ be an independent identically distributed sequence with the same common distribution as the $X_{n,i}$. By the strong law of large numbers, for any $\varepsilon > 0$ the probability that $|Y_1 + \cdots + Y_{2^n}| > \varepsilon 2^n$ infinitely often is 0. Therefore, by the triangle inequality, for any $\varepsilon > 0$ the probability that $|Y_{2^n+1} + \cdots + Y_{2^{n+1}}| > \varepsilon 2^n$ infinitely often is 0; and so, by the Borel–Cantelli lemma for sequences of independent events,

$$\sum_n \mathbb{P}\{|Y_{2^n+1} + \cdots + Y_{2^{n+1}}| > \varepsilon 2^n\} < \infty$$

for all $\varepsilon > 0$. The last sum is also

$$\sum_n \mathbb{P}\{|X_{n,1} + \cdots + X_{n,2^n}| > \varepsilon 2^n\},$$

and an application of the "other half" of the Borel–Cantelli lemma for possibly dependent events establishes that for all $\varepsilon > 0$ the probability of $|X_{n,1} + \cdots + X_{n,2^n}| > \varepsilon 2^n$ infinitely often is 0, as required.   □

We can now give the proof of Theorem 8.2. Proposition 8.8 and Lemma 8.6 verify the conditions of Proposition B.3. The proof is then completed using Equation (10) of [113] that gives upper and lower bounds on the capacity of $C_{\frac{1}{2}}$ in an arbitrary gauge.

Subtree Prune and Re-Graft

9.1 Background

As we mentioned in Chapter 1, Markov chains that move through a space of finite trees are an important ingredient in several algorithms in phylogenetic analysis, and one standard set of moves that is implemented in several phylogenetic software packages is the set of *subtree prune and re-graft* (SPR) moves.

In an SPR move, a binary tree T (that is, a tree in which all non-leaf vertices have degree three) is cut "in the middle of an edge" to give two subtrees, say T' and T''. Another edge is chosen in T', a new vertex is created "in the middle" of that edge, and the cut edge in T'' is attached to this new vertex. Lastly, the "pendant" cut edge in T' is removed along with the vertex it was attached to in order to produce a new binary tree that has the same number of vertices as T – see Figure 9.1.

In this chapter we investigate the asymptotics of the simplest possible tree-valued Markov chain based on the SPR moves, namely the chain in which the two edges that are chosen for cutting and for re-attaching are chosen uniformly (without replacement) from the edges in the current tree. Intuitively, the continuous time Markov process we discuss arises as limit when the number of vertices in the tree goes to infinity, the edge lengths are re-scaled by a constant factor so that initial tree converges in a suitable sense to a continuous analogue of a combinatorial tree (more specifically, a compact real tree), and the time scale of the Markov chain is sped up by an appropriate factor. We do not, in fact, prove such a limit theorem. Rather, we use Dirichlet form techniques to establish the existence of a process that has the dynamics we would expect from such a limit.

The process we construct has as its state space the set of pairs (T, ν), where T is a compact real tree and ν is a probability measure on T. Let μ be the length measure associated with T. Our process jumps away from T by first choosing a pair of points $(u, v) \in T \times T$ according to the rate measure $\mu \otimes \nu$ and then transforming T into a new tree by cutting off the

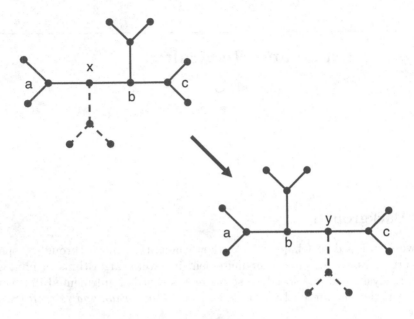

Fig. 9.1. A subtree prune and re-graft operation

subtree rooted at u that does not contain v and re-attaching this subtree at v. This jump kernel (which typically has infinite total mass – so that jumps are occurring on a dense countable set) is precisely what we would expect for a limit (as the number of vertices goes to infinity) of the particular SPR Markov chain on finite trees described above in which the edges for cutting and re-attachment are chosen uniformly at each stage. The limit process is reversible with respect to the distribution of Brownian CRT weighted with the probability measure that comes from the push-forward of Lebesgue measure on $[0, 1]$ as in Example 4.39.

For $\mathbb{R}$-trees arising from an excursion path, the counterpart of an SPR move is the excision and re-insertion of a sub-excursion. Figure 9.2 illustrates such an operation.

We follow the development of [65] in this chapter.

9.2 The Weighted Brownian CRT

Consider the Itô excursion measure for excursions of standard Brownian motion away from 0. This σ-finite measure is defined subject to a normalization of Brownian local time at 0, and we take the usual normalization of local

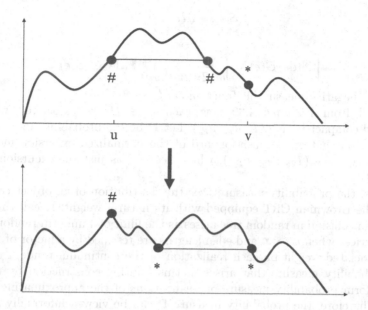

Fig. 9.2. A subtree prune and re-graft operation on an excursion path: the excursion starting at time u in the top picture is excised and inserted at time v, and the resulting gap between the two points marked # is closed up. The two points marked # (resp. *) in the top (resp. bottom) picture correspond to a single point in the associated real tree.

times at each level that makes the local time process an occupation density in the spatial variable for each fixed value of the time variable. The excursion measure is the sum of two measures, one that is concentrated on non-negative excursions and one that is concentrated on non-positive excursions. Let N be the part that is concentrated on non-negative excursions. Thus, in the notation of Example 3.14, N is a σ-finite measure on the space of excursion paths U, where we equip U with the σ-field $\mathcal{U}$ generated by the coordinate maps.

Define a map $v : U \to U^1$ by $e \mapsto \frac{e(\zeta(e)\cdot)}{\sqrt{\zeta(e)}}$. Then

$$\mathbb{P}(\Gamma) := \frac{\mathbb{N}\{v^{-1}(\Gamma) \cap \{e \in U : \zeta(e) \geqslant c\}\}}{\mathbb{N}\{e \in U : \zeta(e) \geqslant c\}}, \quad \Gamma \in \mathcal{U},$$

does not depend on $c > 0$ – see, for example, Exercise 12.2.13.2 in [117]. The probability measure $\mathbb{P}$ is called the law of normalized non-negative Brownian excursion. We have

$$\mathbb{N}\{e \in U : \zeta(e) \in dc\} = \frac{dc}{2\sqrt{2\pi c^3}} \tag{9.1}$$

and, defining $\mathcal{S}_c : U^1 \to U^c$ by

$$\mathcal{S}_c e := \sqrt{c}e(\cdot/c) \tag{9.2}$$

we have

$$\int \mathbb{N}(\mathrm{d}e)\,G(e) = \int_0^\infty \frac{\mathrm{d}c}{2\sqrt{2\pi c^3}} \int_{U^1} \mathbb{P}(\mathrm{d}e)\,G\left(\mathcal{S}_c e\right) \tag{9.3}$$

for a non-negative measurable function $G : U \to \mathbb{R}$.

Recall from Example 4.39 how each $e \in U^1$ is associated with a weighted compact $\mathbb{R}$-tree $(T_e, d_{T_e}, \nu_{T_e})$. Let $\mathbf{P}$ be the probability measure on $(\mathbf{T}^{\mathrm{wt}}, d_{\mathrm{GH}^{\mathrm{wt}}})$ that is the push-forward of the normalized excursion measure by the map $e \mapsto (T_{2e}, d_{T_{2e}}, \nu_{T_{2e}})$, where $2e \in U^1$ is just the excursion path $t \mapsto 2e(t)$.

Thus, the probability measure $\mathbf{P}$ is the distribution of an object consisting of the Brownian CRT equipped with its natural weight. Recall that the Brownian continuum random tree arises as the limit of a uniform random tree on n vertices when $n \to \infty$ and edge lengths are rescaled by a factor of $1/\sqrt{n}$. The associated weight on each realization of the continuum random tree is the probability measure that arises in this limiting construction by taking the uniform probability measure on realizations of the approximating finite trees. Therefore, the probability measure $\mathbf{P}$ can be viewed informally as the "uniform distribution" on $(\mathbf{T}^{\mathrm{wt}}, d_{\mathrm{GH}^{\mathrm{wt}}})$.

9.3 Campbell Measure Facts

For the purposes of constructing the Markov process that is of interest to us, we need to understand picking a random weighted tree (T, d_T, ν_T) according to the continuum random tree distribution $\mathbf{P}$, picking a point u according to the length measure μ^T and another point v according to the weight ν_T, and then decomposing T into two subtrees rooted at u – one that contains v and one that does not (we are being a little imprecise here, because μ^T will be an infinite measure, $\mathbf{P}$ almost surely).

In order to understand this decomposition, we must understand the corresponding decomposition of excursion paths under normalized excursion measure. Because subtrees correspond to sub-excursions and because of our observation in Example 4.34 that for an excursion e the length measure μ^{T_e} on the corresponding tree is the push-forward of the measure $\int_{\Gamma_e} \mathrm{d}s \otimes \mathrm{d}a\, \frac{1}{\bar{s}(e,s,a) - \underline{s}(e,s,a)} \delta_{\underline{s}(e,s,a)}$ by the quotient map, we need to understand the decomposition of the excursion e into the excursion above a that straddles s and the "remaining" excursion when when e is chosen according to the standard Brownian excursion distribution $\mathbb{P}$ and (s, a) is chosen according to the σ-finite measure $\mathrm{d}s \otimes \mathrm{d}a\, \frac{1}{\bar{s}(e,s,a) - \underline{s}(e,s,a)}$ on Γ_e – see Figure 9.3.

Given an excursion $e \in U$ and a level $a \geqslant 0$ write:

- $\zeta(e) := \inf\{t > 0 : e(t) = 0\}$ for the "length" of e,

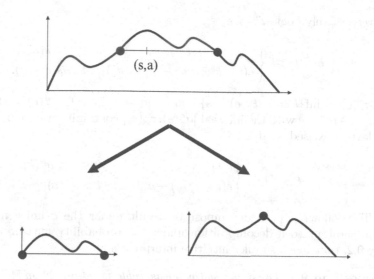

Fig. 9.3. The decomposition of the excursion e in the top picture into the excursion $\hat{e}^{s,a}$ above level a that straddles time s in the middle picture and the "remaining" excursion $\check{e}^{s,a}$ in the bottom picture.

- $\ell_t^a(e)$ for the local time of e at level a up to time t,
- $e^{\downarrow a}$ for e time-changed by the inverse of $t \mapsto \int_0^t ds\, 1\{e(s) \leqslant a\}$ (that is, $e^{\downarrow a}$ is e with the sub-excursions above level a excised and the gaps closed up),
- $\ell_t^a(e^{\downarrow a})$ for the local time of $e^{\downarrow a}$ at the level a up to time t,
- $U^{\uparrow a}(e)$ for the set of sub-excursion intervals of e above a (that is, an element of $U^{\uparrow a}(e)$ is an interval $I = [g_I, d_I]$ such that $e(g_I) = e(d_I) = a$ and $e(t) > a$ for $g_I < t < d_I$),
- $\mathcal{N}^{\uparrow a}(e)$ for the counting measure that puts a unit mass at each point (s', e'), where, for some $I \in U^{\uparrow a}(e)$, $s' := \ell_{g_I}^a(e)$ is the amount of local time of e at level a accumulated up to the beginning of the sub-excursion I and $e' \in U$ is given by

$$
e'(t) = \begin{cases} e(g_I + t) - a, & 0 \leqslant t \leqslant d_I - g_I, \\ 0, & t > d_I - g_I, \end{cases}
$$

  is the corresponding piece of the path e shifted to become an excursion above the level 0 starting at time 0,
- $\hat{e}^{s,a} \in U$ and $\check{e}^{s,a} \in U$, for the subexcursion "above" $(s, a) \in \Gamma_e$, that is,

$$\hat{e}^{s,a}(t) := \begin{cases} e(\underline{s}(e,s,a) + t) - a, & 0 \leqslant t \leqslant \bar{s}(e,s,a) - \underline{s}(e,s,a), \\ 0, & t > \bar{s}(e,s,a) - \underline{s}(e,s,a), \end{cases}$$

respectively "below" $(s,a) \in \Gamma_e$, that is,

$$\check{e}^{s,a}(t) := \begin{cases} e(t), & 0 \leqslant t \leqslant \underline{s}(e,s,a), \\ e(t + \bar{s}(e,s,a) - \underline{s}(e,s,a)), & t > \underline{s}(e,s,a). \end{cases}$$

- $\sigma_s^a(e) := \inf\{t \geqslant 0 : \ell_t^a(e) \geqslant s\}$ and $\tau_s^a(e) := \inf\{t \geqslant 0 : \ell_t^a(e) > s\}$,
- $\tilde{e}^{s,a} \in U$ for e with the interval $]\sigma_s^a(e), \tau_s^a(e)[$ containing an excursion above level a excised, that is,

$$\tilde{e}^{s,a}(t) := \begin{cases} e(t), & 0 \leqslant t \leqslant \sigma_s^a(e), \\ e(t + \tau_s^a(e) - \sigma_s^a(e)), & t > \sigma_s^a(e). \end{cases}$$

The following path decomposition result under the σ-finite measure $\mathbb{N}$ is preparatory to a decomposition under the probability measure $\mathbb{P}$, Corollary 9.2, that has a simpler intuitive interpretation.

Proposition 9.1. *For non-negative measurable functions F on $\mathbb{R}_+$ and G, H on U,*

$$\int \mathbb{N}(de) \int_{\Gamma_e} \frac{ds \otimes da}{\bar{s}(e,s,a) - \underline{s}(e,s,a)} F(\underline{s}(e,s,a)) G(\hat{e}^{s,a}) H(\check{e}^{s,a})$$

$$= \int \mathbb{N}(de) \int_0^\infty da \int \mathcal{N}^{\uparrow a}(e)(d(s',e')) F(\sigma_{s'}^a(e)) G(e') H(\tilde{e}^{s',a})$$

$$= \mathbb{N}[G] \, \mathbb{N}\Big[H \int_0^\zeta ds \, F(s)\Big].$$

Proof. The first equality is just a change in the order of integration and has already been remarked upon in Example 4.34.

Standard excursion theory – see, for example, [119, 117, 29] – says that under $\mathbb{N}$, the random measure $e \mapsto \mathcal{N}^{\uparrow a}(e)$ conditional on $e \mapsto e^{\downarrow a}$ is a Poisson random measure with intensity measure $\lambda^{\downarrow a}(e) \otimes \mathbb{N}$, where $\lambda^{\downarrow a}(e)$ is Lebesgue measure restricted to the interval $[0, \ell_\infty^a(e)] = [0, 2\ell_\infty^a(e^{\downarrow a})]$.

Note that $\tilde{e}^{s',a}$ is constructed from $e^{\downarrow a}$ and $\mathcal{N}^{\uparrow a}(e) - \delta_{(s',e')}$ in the same way that e is constructed from $e^{\downarrow a}$ and $\mathcal{N}^{\uparrow a}(e)$. Also, $\sigma_{s'}^a(\tilde{e}^{s',a}) = \sigma_{s'}^a(e)$. Therefore, by the Campbell–Palm formula for Poisson random measures – see, for example, Section 12.1 of [41] –

$$\int N(de) \int_0^\infty da \int \mathcal{N}^{\uparrow a}(e)(d(s',e')) \, F(\sigma_{s'}^a(e))G(e')H(\tilde{e}^{s',a})$$

$$= \int N(de) \int_0^\infty da \, N \Big[\int \mathcal{N}^{\uparrow a}(e)(d(s',e')) \, F(\sigma_{s'}^a(e))G(e')H(\tilde{e}^{s',a}) \Big| e^{\downarrow a} \Big]$$

$$= \int N(de) \int_0^\infty da \, N[G] \, N \Big[\Big\{ \int_0^{\ell_\infty^a(e)} ds' \, F(\sigma_{s'}^a(e)) \Big\} H \Big| e^{\downarrow a} \Big]$$

$$= N[G] \int_0^\infty da \int N(de) \left(\Big\{ \int d\ell_s^a(e) \, F(s) \Big\} H(e) \right)$$

$$= N[G] \int N(de) \left(\Big\{ \int_0^\infty da \int d\ell_s^a(e) \, F(s) \Big\} H(e) \right)$$

$$= N[G] \, N \Big[H \int_0^\zeta ds \, F(s) \Big].$$

□

The next result says that if we pick an excursion e according to the standard excursion distribution $\mathbb{P}$ and then pick a point $(s,a) \in \Gamma_e$ according to the σ-finite length measure corresponding to the length measure μ^{T_e} on the associated tree T_e, then the following objects are independent:

(a) the length of the excursion above level a that straddles time s,
(b) the excursion obtained by taking the excursion above level a that straddles time s, turning it (by a shift of axes) into an excursion $\hat{e}^{s,a}$ above level zero starting at time zero, and then Brownian re-scaling $\hat{e}^{s,a}$ to produce an excursion of unit length,
(c) the excursion obtained by taking the excursion $\check{e}^{s,a}$ that comes from excising $\hat{e}^{s,a}$ and closing up the gap, and then Brownian re-scaling $\check{e}^{s,a}$ to produce an excursion of unit length,
(d) the starting time $\underline{s}(e,s,a)$ of the excursion above level a that straddles time s rescaled by the length of $\check{e}^{s,a}$ to give a time in the interval $[0,1]$.

Moreover, the length in (a) is "distributed" according to the σ-finite measure

$$\frac{1}{2\sqrt{2\pi}} \frac{d\rho}{\sqrt{(1-\rho)\rho^3}}, \quad 0 \le \rho \le 1,$$

the unit length excursions in (b) and (c) are both distributed as standard Brownian excursions (that is, according to $\mathbb{P}$), and the time in (d) is uniformly distributed on the interval $[0,1]$.

Corollary 9.2. *For non-negative measurable functions F on $\mathbb{R}_+$ and K on $U \times U$,*

$$\int \mathbb{P}(de) \int_{\Gamma_e} \frac{ds \otimes da}{\bar{s}(e,s,a) - \underline{s}(e,s,a)} F\left(\frac{\underline{s}(e,s,a)}{\zeta(\check{e}^{s,a})}\right) K(\hat{e}^{s,a}, \check{e}^{s,a})$$

$$= \left\{ \int_0^1 du\, F(u) \right\} \int \mathbb{P}(de) \int_{\Gamma_e} \frac{ds \otimes da}{\bar{s}(e,s,a) - \underline{s}(e,s,a)} K(\hat{e}^{s,a}, \check{e}^{s,a})$$

$$= \left\{ \int_0^1 du\, F(u) \right\} \frac{1}{2\sqrt{2\pi}} \int_0^1 \frac{d\rho}{\sqrt{(1-\rho)\rho^3}} \int \mathbb{P}(de') \otimes \mathbb{P}(de'')\, K(\mathcal{S}_\rho e', \mathcal{S}_{1-\rho} e'').$$

Proof. For a non-negative measurable function L on $U \times U$, it follows straightforwardly from Proposition 9.1 that

$$\int \mathbb{N}(de) \int_{\Gamma_e} \frac{ds \otimes da}{\bar{s}(e,s,a) - \underline{s}(e,s,a)} F\left(\frac{\underline{s}(e,s,a)}{\zeta(\check{e}^{s,a})}\right) L(\hat{e}^{s,a}, \check{e}^{s,a})$$

$$= \left\{ \int_0^1 du\, F(u) \right\} \int \mathbb{N}(de') \otimes \mathbb{N}(de'')\, L(e', e'')\zeta(e''). \qquad (9.4)$$

The left-hand side of equation (9.4) is, by (9.3),

$$\int_0^\infty \frac{dc}{2\sqrt{2\pi c^3}} \int \mathbb{P}(de) \int_{\Gamma_{\mathcal{S}_c e}} ds \otimes da\, \frac{F\left(\frac{\underline{s}(\mathcal{S}_c e, s, a)}{\zeta(\widehat{\mathcal{S}_c e}^{s,a})}\right) L(\widehat{\mathcal{S}_c e}^{s,a}, \widecheck{\mathcal{S}_c e}^{s,a})}{\bar{s}(\mathcal{S}_c e, s, a) - \underline{s}(\mathcal{S}_c e, s, a)}. \qquad (9.5)$$

If we change variables to $t = s/c$ and $b = a/\sqrt{c}$, then the integral for (s,a) over $\Gamma_{\mathcal{S}_c e}$ becomes an integral for (t,b) over Γ_e. Also,

$$\underline{s}(\mathcal{S}_c e, ct, \sqrt{c}b) = \sup\left\{ r < ct : \sqrt{c}e\left(\frac{r}{c}\right) < \sqrt{c}b \right\}$$

$$= c\sup\left\{ r < t : e(r) < b \right\}$$

$$= c\underline{s}(e, t, b),$$

and, by similar reasoning,

$$\bar{s}(\mathcal{S}_c e, ct, \sqrt{c}b) = c\bar{s}(e, t, b)$$

and

$$\zeta(\widehat{\mathcal{S}_c e}^{ct, \sqrt{c}b}) = c\zeta(\check{e}^{t,b}).$$

Thus, (9.5) is

$$\int_0^\infty \frac{dc}{2\sqrt{2\pi c^3}} \int \mathbb{P}(de) \sqrt{c} \int_{\Gamma_e} dt \otimes db\, \frac{F\left(\frac{\underline{s}(e,t,b)}{\zeta(\check{e}^{t,b})}\right) L(\widehat{\mathcal{S}_c e}^{ct, \sqrt{c}b}, \widecheck{\mathcal{S}_c e}^{ct, \sqrt{c}b})}{\bar{s}(e,t,b) - \underline{s}(e,t,b)}. \qquad (9.6)$$

Now suppose that L is of the form

$$L(e', e'') = K(\mathcal{R}_{\zeta(e')+\zeta(e'')} e', \mathcal{R}_{\zeta(e')+\zeta(e'')} e'') \frac{M(\zeta(e') + \zeta(e''))}{\sqrt{\zeta(e') + \zeta(e'')}},$$

where, for ease of notation, we put for $e \in U$, and $c > 0$,

$$\mathcal{R}_c e := \mathcal{S}_{c^{-1}} e = \frac{1}{\sqrt{c}} e(c \cdot).$$

Then (9.6) becomes

$$\int_0^\infty \frac{dc}{2\sqrt{2\pi c^3}} \int \mathbb{P}(de) \int_{\Gamma_e} dt \otimes db \, \frac{F\left(\frac{s(e,t,b)}{\zeta(\breve{e}^{t,b})}\right) K(\hat{e}^{t,b}, \breve{e}^{t,b}) M(c)}{\bar{s}(e,t,b) - \underline{s}(e,t,b)}. \tag{9.7}$$

Since (9.7) was shown to be equivalent to the left hand side of (9.4), it follows from (9.3) that

$$\int \mathbb{P}(de) \int_{\Gamma_e} \frac{dt \otimes db}{\bar{s}(e,t,b) - \underline{s}(e,t,b)} F\left(\frac{s(e,t,b)}{\zeta(\breve{e}^{t,b})}\right) K(\hat{e}^{t,b}, \breve{e}^{t,b})$$
$$= \frac{\int_0^1 du \, F(u)}{N[M]} \int \mathbb{N}(de') \otimes \mathbb{N}(de'') L(e', e'') \zeta(e''), \tag{9.8}$$

and the first equality of the statement follows.

We have from the identity (9.8) that, for any $C > 0$,

$$N\{\zeta(e) > C\} \int \mathbb{P}(de) \int_{\Gamma_e} \frac{ds \otimes da}{\bar{s}(e,s,a) - \underline{s}(e,s,a)} K(\hat{e}^{s,a}, \breve{e}^{s,a})$$

$$= \int \mathbb{N}(de') \otimes \mathbb{N}(de'') K(\mathcal{R}_{\zeta(e')+\zeta(e'')} e', \mathcal{R}_{\zeta(e')+\zeta(e'')} e'') \frac{1\{\zeta(e') + \zeta(e'') > C\}}{\sqrt{\zeta(e') + \zeta(e'')}} \zeta(e'')$$

$$= \int_0^\infty \frac{dc'}{2\sqrt{2\pi c'^3}} \int_0^\infty \frac{dc''}{2\sqrt{2\pi c''}}$$
$$\int \mathbb{P}(de') \otimes \mathbb{P}(de'') K(\mathcal{R}_{c'+c''} \mathcal{S}_{c'} e', \mathcal{R}_{c'+c''} \mathcal{S}_{c''} e'') \frac{1\{c' + c'' > C\}}{\sqrt{c' + c''}}.$$

Make the change of variables $\rho = \frac{c'}{c'+c''}$ and $\xi = c' + c''$ (with corresponding Jacobian factor ξ) to get

$$\int_0^\infty \frac{dc'}{2\sqrt{2\pi c'^3}} \int_0^\infty \frac{dc''}{2\sqrt{2\pi c''}}$$
$$\int \mathbb{P}(de') \otimes \mathbb{P}(de'') K(\mathcal{R}_{c'+c''} \mathcal{S}_{c'} e', \mathcal{R}_{c'+c''} \mathcal{S}_{c''} e'') \frac{1\{c' + c'' > C\}}{\sqrt{c' + c''}}$$

$$= \left(\frac{1}{2\sqrt{2\pi}}\right)^2 \int_0^\infty d\xi \int_0^1 \frac{d\rho \, \xi}{\sqrt{\rho^3(1-\rho)\xi^4}} \frac{1\{\xi > C\}}{\sqrt{\xi}}$$
$$\int \mathbb{P}(de') \otimes \mathbb{P}(de'') K(\mathcal{S}_\rho e', \mathcal{S}_{1-\rho} e'')$$

$$= \left(\frac{1}{2\sqrt{2\pi}}\right)^2 \left\{\int_C^\infty \frac{d\xi}{\sqrt{\xi^3}}\right\} \int_0^1 \frac{d\rho}{\sqrt{\rho^3(1-\rho)}}$$
$$\int \mathbb{P}(de') \otimes \mathbb{P}(de'') K(\mathcal{S}_\rho e', \mathcal{S}_{1-\rho} e''),$$

and the corollary follows upon recalling (9.1). □

Corollary 9.3. *(i) For $x > 0$,*

$$\int \mathbb{P}(de) \int_{\Gamma_e} \frac{ds \otimes da}{\bar{s}(e,s,a) - \underline{s}(e,s,a)} 1\{\max_{0 \leqslant t \leqslant \zeta(\hat{e}^{s,a})} \hat{e}^{s,a} > x\}$$

$$= 2 \sum_{n \in \mathbb{N}} nx \, \exp(-2n^2 x^2)$$

(ii) For $0 < p \leqslant 1$,

$$\int \mathbb{P}(de) \int_{\Gamma_e} \frac{ds \otimes da}{\bar{s}(e,s,a) - \underline{s}(e,s,a)} 1\{\zeta(\hat{e}^{s,a}) > p\} = \sqrt{\frac{1-p}{2\pi p}}.$$

Proof. (i) Recall first of all from Theorem 5.2.10 in [92] that

$$\mathbb{P}\left\{e \in U^1 : \max_{0 \leqslant t \leqslant 1} e(t) > x\right\} = 2 \sum_{n \in \mathbb{N}} (4n^2 x^2 - 1) \exp(-2n^2 x^2).$$

By Corollary 9.2 applied to $K(e', e'') := 1\{\max_{t \in [0, \zeta(e')]} e'(t) \geqslant x\}$ and $F \equiv 1$,

$$\int \mathbb{P}(de) \int_{\Gamma_e} \frac{ds \otimes da}{\bar{s}(e,s,a) - \underline{s}(e,s,a)} 1\{\max_{0 \leqslant t \leqslant \zeta(\hat{e}^{s,a})} \hat{e}^{s,a} > x\}$$

$$= \frac{1}{2\sqrt{2\pi}} \int_0^1 \frac{d\rho}{\sqrt{\rho^3(1-\rho)}} \mathbb{P}\left\{\max_{t \in [0,\rho]} \sqrt{\rho} e(t/\rho) > x\right\}$$

$$= \frac{1}{2\sqrt{2\pi}} \int_0^1 \frac{d\rho}{\sqrt{\rho^3(1-\rho)}} \mathbb{P}\left\{\max_{t \in [0,1]} e(t) > \frac{x}{\sqrt{\rho}}\right\}$$

$$= \frac{1}{2\sqrt{2\pi}} \int_0^1 \frac{d\rho}{\sqrt{\rho^3(1-\rho)}} 2 \sum_{n \in \mathbb{N}} \left(4n^2 \frac{x^2}{\rho} - 1\right) \exp\left(-2n^2 \frac{x^2}{\rho}\right)$$

$$= 2 \sum_{n \in \mathbb{N}} nx \, \exp(-2n^2 x^2),$$

as claimed.

(ii) Corollary 9.2 applied to $K(e', e'') := 1\{\zeta(e') \geqslant p\}$ and $F \equiv 1$ immediately yields

$$\int \mathbb{P}(de) \int_{\Gamma_e} \frac{ds \otimes da}{\bar{s}(e,s,a) - \underline{s}(e,s,a)} 1\{\zeta(\hat{e}^{s,a}) > p\}$$

$$= \frac{1}{2\sqrt{2\pi}} \int_p^1 \frac{d\rho}{\sqrt{\rho^3(1-\rho)}} = \sqrt{\frac{1-p}{2\pi p}}.$$

$\square$

We conclude this section by calculating the expectations of some functionals with respect to **P** (the the "uniform distribution" on $(\mathbf{T}^{\mathrm{wt}}, d_{\mathrm{GH}^{\mathrm{wt}}})$ as introduced in the end of Section 9.2).

For $\varepsilon > 0$, $T \in \mathbf{T}$, and $\rho \in T$, write $R_\varepsilon(T, \rho)$ for the ε-trimming of the rooted $\mathbb{R}$-tree obtained by rooting T at ρ (recall Subsection 4.3.4). With a slight abuse of notation, set

$$R_\varepsilon(T) := \begin{cases} \bigcap_{\rho \in T} R_\varepsilon(T, \rho), & \operatorname{diam}(T) > \varepsilon, \\ \text{singleton}, & \operatorname{diam}(T) \leqslant \varepsilon. \end{cases} \tag{9.9}$$

For $T \in \mathbf{T}^{\mathrm{wt}}$ recall the length measure μ^T from (4.10). Given $(T, d) \in \mathbf{T}^{\mathrm{wt}}$ and $u, v \in T$, let

$$S^{T,u,v} := \{x \in T : u \in]v, x[\}, \tag{9.10}$$

denote the subtree of T that differs from its closure by the point u, which can be thought of as its root, and consists of points that are on the "other side" of u from v (recall $]v, x[$ is the open segment in T between v and x).

Lemma 9.4. *(i) For $x > 0$,*

$$\mathbf{P}\left[\mu^T \otimes \nu_T \left\{(u, v) \in T \times T : \operatorname{height}(S^{T,u,v}) > x\right\}\right]$$

$$= \mathbf{P}\left[\int_T \nu_T(\mathrm{d}v)\, \mu^T(R_x(T, v))\right]$$

$$= 2 \sum_{n \in \mathbb{N}} nx \, \exp(-n^2 x^2 / 2).$$

(ii) For $1 < \alpha < \infty$,

$$\mathbf{P}\left[\int_T \nu_T(\mathrm{d}v) \int_T \mu^T(\mathrm{d}u) \left(\operatorname{height}(S^{T,u,v})\right)^\alpha\right]$$

$$= 2^{-\frac{1}{2}}\, \alpha\, \Gamma\!\left(\frac{\alpha}{2} + \frac{1}{2}\right) \zeta(\alpha),$$

where, as usual, $\zeta(\alpha) := \sum_{n \geqslant 1} n^{-\alpha}$.
(iii) For $0 < p \leqslant 1$,

$$\mathbf{P}\left[\mu^T \otimes \nu_T\{(u, v) \in T \times T : \nu_T(S^{T,u,v}) > p\}\right] = \sqrt{\frac{2(1 - p)}{\pi p}}.$$

(iv) For $\frac{1}{2} < \beta < \infty$,

$$\mathbf{P}\left[\int_T \nu_T(\mathrm{d}v) \int_T \mu^T(\mathrm{d}u) \left(\nu_T\left(S^{T,u,v}\right)\right)^\beta\right] = 2^{-\frac{1}{2}} \frac{\Gamma\left(\beta - \frac{1}{2}\right)}{\Gamma(\beta)}.$$

Proof. (i) The first equality is clear from the definition of $R_x(T, v)$ and Fubini's theorem.

Turning to the equality of the first and last terms, first recall that $\mathbf{P}$ is the push-forward on $(\mathbf{T}^{\mathrm{wt}}, d_{\mathrm{GH}^{\mathrm{wt}}})$ of the normalized excursion measure $\mathbb{P}$ by the map $e \mapsto (T_{2e}, d_{T_{2e}}, \nu_{T_{2e}})$, where $2e \in U^1$ is just the excursion path $t \mapsto 2e(t)$. In particular, T_{2e} is the quotient of the interval $[0, 1]$ by the equivalence relation defined by $2e$. By the invariance of the standard Brownian excursion under random re-rooting – see Section 2.7 of [13] – the point in T_{2e} that corresponds to the equivalence class of $0 \in [0, 1]$ is distributed according to

$\nu_{T_{2e}}$ when e is chosen according to $\mathbb{P}$. Moreover, recall from Example 4.34 that for $e \in U^1$, the length measure μ^{T_e} is the push-forward of the measure $\mathrm{d}s \otimes \mathrm{d}a \frac{1}{\overline{s}(e,s,a) - \underline{s}(e,s,a)} \delta_{\underline{s}(e,s,a)}$ on the sub-graph Γ_e by the quotient map defined in (3.14).

It follows that if we pick T according to $\mathbf{P}$ and then pick $(u,v) \in T \times T$ according to $\mu^T \otimes \nu_T$, then the subtree $S^{T,u,v}$ that arises has the same σ-finite law as the tree associated with the excursion $2\hat{e}^{s,a}$ when e is chosen according to $\mathbb{P}$ and (s,a) is chosen according to the measure $\mathrm{d}s \otimes \mathrm{d}a \frac{1}{\overline{s}(e,s,a) - \underline{s}(e,s,a)} \delta_{\underline{s}(e,s,a)}$ on the sub-graph Γ_e.

Therefore, by part (i) of Corollary 9.3,

$$\mathbf{P}\left[\int_T \nu_T(\mathrm{d}v) \int_T \mu^T(\mathrm{d}u) 1\left\{ \mathrm{height}(S^{T,u,v}) > x \right\} \right]$$

$$= 2 \int \mathbb{P}(\mathrm{d}e) \int_{\Gamma_e} \frac{\mathrm{d}s \otimes \mathrm{d}a}{\overline{s}(e,s,a) - \underline{s}(e,s,a)} 1\left\{ \max_{0 \leqslant t \leqslant \zeta(\hat{e}^{s,a})} \hat{e}^{s,a} > \frac{x}{2} \right\}$$

$$= 2 \sum_{n \in \mathbb{N}} nx \exp(-n^2 x^2 / 2).$$

Part (ii) is a consequence of part (i) and some straightforward calculus.

Part (iii) follows immediately from part(ii) of Corollary 9.3.

Part (iv) is a consequence of part (iii) and some more straightforward calculus.   □

9.4 A Symmetric Jump Measure

In this section we will construct and study a measure on $\mathbf{T}^{\mathrm{wt}} \times \mathbf{T}^{\mathrm{wt}}$ that is related to the decomposition discussed at the beginning of Section 9.3.

Define a map Θ from $\{((T,d),u,v) : T \in \mathbf{T}, u \in T, v \in T\}$ into $\mathbf{T}$ by setting $\Theta((T,d),u,v) := (T, d^{(u,v)})$ where letting

$$d^{(u,v)}(x,y) := \begin{cases} d(x,y), & \text{if } x, y \in S^{T,u,v}, \\ d(x,y), & \text{if } x, y \in T \backslash S^{T,u,v}, \\ d(x,u) + d(v,y), & \text{if } x \in S^{T,u,v}, y \in T \backslash S^{T,u,v}, \\ d(y,u) + d(v,x), & \text{if } y \in S^{T,u,v}, x \in T \backslash S^{T,u,v}. \end{cases}$$

That is, $\Theta((T,d),u,v)$ is just T as a set, but the metric has been changed so that the subtree $S^{T,u,v}$ with root u is now pruned and re-grafted so as to have root v.

If $(T,d,\nu) \in \mathbf{T}^{\mathrm{wt}}$ and $(u,v) \in T \times T$, then we can think of ν as a weight on $(T, d^{(u,v)})$, because the Borel structures induces by d and $d^{(u,v)}$ are the same. With a slight misuse of notation we will, therefore, write $\Theta((T,d,\nu),u,v)$ for $(T, d^{(u,v)}, \nu) \in \mathbf{T}^{\mathrm{wt}}$. Intuitively, the mass contained in $S^{T,u,v}$ is transported along with the subtree.

Define a kernel κ on $\mathbf{T}^{\mathrm{wt}}$ by

$$\kappa((T, d_T, \nu_T), \mathbf{B}) := \mu^T \otimes \nu_T \{(u, v) \in T \times T : \Theta(T, u, v) \in \mathbf{B}\}$$

for $\mathbf{B} \in \mathcal{B}(\mathbf{T}^{\mathrm{wt}})$. Thus, $\kappa((T, d_T, \nu_T), \cdot)$ is the jump kernel described informally in Section 9.1.

We show in part (i) of Lemma 9.5 below that the kernel κ is reversible with respect to the probability measure $\mathbf{P}$. More precisely, we show that if we define a measure J on $\mathbf{T}^{\mathrm{wt}} \times \mathbf{T}^{\mathrm{wt}}$ by

$$J(\mathbf{A} \times \mathbf{B}) := \int_{\mathbf{A}} \mathbf{P}(\mathrm{d}T) \, \kappa(T, \mathbf{B})$$

for $\mathbf{A}, \mathbf{B} \in \mathcal{B}(\mathbf{T}^{\mathrm{wt}})$, then J is symmetric.

Lemma 9.5. *Then*

(i) The measure J is symmetric.

(ii) For each compact subset $\mathbf{K} \subset \mathbf{T}^{\mathrm{wt}}$ and open subset $\mathbf{U}$ such that $\mathbf{K} \subset \mathbf{U} \subseteq \mathbf{T}$,

$$J(\mathbf{K}, \mathbf{T}^{\mathrm{wt}} \backslash \mathbf{U}) < \infty.$$

(iii)

$$\int_{\mathbf{T}^{\mathrm{wt}} \times \mathbf{T}^{\mathrm{wt}}} J(\mathrm{d}T, \mathrm{d}S) \, \Delta_{\mathrm{GH}^{\mathrm{wt}}}^2 (T, S) < \infty.$$

Proof. (i) Given $e', e'' \in U^1, 0 \leqslant u \leqslant 1$, and $0 < \rho \leqslant 1$, define $e^\circ(\cdot; e', e'', u, \rho) \in U^1$ by

$e^\circ(t; e', e'', u, \rho)$

$$:= \begin{cases} \mathcal{S}_{1-\rho} e''(t), & 0 \leqslant t \leqslant (1-\rho)u, \\ \mathcal{S}_{1-\rho} e''((1-\rho)u) + \mathcal{S}_\rho e'(t - (1-\rho)u), & (1-\rho)u \leqslant t \leqslant (1-\rho)u + \rho, \\ \mathcal{S}_{1-\rho} e''(t - \rho), & (1-\rho)u + \rho \leqslant t \leqslant 1. \end{cases}$$

That is, $e^\circ(\cdot; e', e'', u, \rho)$ is the excursion that arises from Brownian re-scaling e' and e'' to have lengths ρ and $1 - \rho$, respectively, and then inserting the re-scaled version of e' into the re-scaled version of e'' at a position that is a fraction u of the total length of the re-scaled version of e''.

Define a measure $\mathbb{J}$ on $U^1 \times U^1$ by

$$\int_{U^1 \times U^1} \mathbb{J}(\mathrm{d}e^*, \mathrm{d}e^{**}) K(e^*, e^{**})$$

$$:= \int_{[0,1]^2} \mathrm{d}u \otimes \mathrm{d}v \frac{1}{2\sqrt{2\pi}} \int_0^1 \frac{\mathrm{d}\rho}{\sqrt{(1-\rho)\rho^3}} \int \mathbb{P}(\mathrm{d}e') \otimes \mathbb{P}(\mathrm{d}e'')$$
$$\times K\left(e^\circ(\cdot; e', e'', u, \rho), e^\circ(\cdot; e', e'', v, \rho)\right).$$

Clearly, the measure $\mathbb{J}$ is symmetric. It follows from the discussion at the beginning of the proof of part (i) of Lemma 9.4 and Corollary 9.2 that the measure J is the push-forward of the symmetric measure $2\mathbb{J}$ by the map that sends the pair $(e^*, e^{**}) \in U^1 \times U^1$ to the pair

$$((T_{2e^*}, d_{T_{2e^*}}, \nu_{T_{2e^*}}), (T_{2e^{**}}, d_{T_{2e^{**}}}, \nu_{T_{2e^{**}}})).$$

Hence, J is also symmetric.

(ii) The result is trivial if $\mathbf{K} = \varnothing$, so we assume that $\mathbf{K} \neq \varnothing$. Since $\mathbf{T}^{\text{wt}} \backslash \mathbf{U}$ and $\mathbf{K}$ are disjoint closed sets and $\mathbf{K}$ is compact, we have that

$$c := \inf_{T \in \mathbf{K}, S \in \mathbf{U}} \Delta_{\text{GH}^{\text{wt}}}(T, S) > 0.$$

Fix $T \in \mathbf{K}$. If $(u, v) \in T \times T$ is such that $\Delta_{\text{GH}}(T, \Theta(T, u, v)) \geqslant c$, then either

- $u \in R_c(T)$, or
- there exists $\rho \in T^o$ such that $u \notin R_c(T, \rho)$ and $\nu_T(S^{T, u, \rho}) \geqslant c$ (recall that $R_c(T)$ is the c-trimming of T, that $R_c(T, \rho)$ is the c-trimming of T rooted at ρ, and that $S^{T, u, \rho}$ is the subtree of T consisting of points that are on the other side of u to ρ).

Hence, we have

$$J(\mathbf{K}, \mathbf{T}^{\text{wt}} \backslash \mathbf{U})$$
$$\leqslant \int_{\mathbf{K}} \mathbf{P}\{dT\} \kappa(T, \{S : \Delta_{\text{GH}^{\text{wt}}}(T, S) > c\})$$
$$\leqslant \int_{\mathbf{K}} \mathbf{P}(dT) \mu^T(R_c(T))$$
$$+ \int_{\mathbf{K}} \mathbf{P}(dT) \int_T \nu_T(dv) \mu^T \{u \in T : \nu_T(S^{T, u, v}) > c\}$$
$$< \infty,$$

where we have used Lemma 9.4 and the observation that

$$\mu^T(R_c(T)) \leqslant \int_T \nu_T(dv) \mu^T(R_c(T, v))$$

because $R_c(T) \subseteq R_c(T, v)$ for all $v \in T$.

(iii) Similar reasoning yields that

$$\int_{\mathbf{T}^{\text{wt}} \times \mathbf{T}^{\text{wt}}} J(\mathrm{d}T, \mathrm{d}S) \, \Delta^2_{\text{GH}^{\text{wt}}} (T, S)$$

$$= \int_{\mathbf{T}^{\text{wt}}} \mathbf{P}\{\mathrm{d}T\} \int_0^\infty \mathrm{d}t \, 2t \, \kappa(T, \{S : \Delta_{\text{GH}^{\text{wt}}}(T, S) > t\})$$

$$\leqslant \int_{\mathbf{T}^{\text{wt}}} \mathbf{P}(\mathrm{d}T) \int_0^\infty \mathrm{d}t \, 2t \, \mu^T (R_t(T))$$

$$\qquad + \int_{\mathbf{T}^{\text{wt}}} \mathbf{P}(\mathrm{d}T) \int_0^\infty \mathrm{d}t \, 2t \int_T \nu_T(\mathrm{d}v) \mu^T \{u \in T : \nu_T\{S^{T,u,v}\} > t\}$$

$$\leqslant \int_0^\infty \mathrm{d}t \, 2t \int_{\mathbf{T}^{\text{wt}}} \mathbf{P}(\mathrm{d}T) \mu^T (R_t(T))$$

$$\qquad + \int_{\mathbf{T}^{\text{wt}}} \mathbf{P}(\mathrm{d}T) \int_T \nu_T(\mathrm{d}v) \int_T \mu^T(\mathrm{d}u) \nu_T^2(S^{T,u,v})$$

$$< \infty,$$

where we have applied Lemma 9.4 once more. $\qquad\qquad\square$

9.5 The Dirichlet Form

Consider the bilinear form

$$\mathcal{E}(f, g) := \int_{\mathbf{T}^{\text{wt}} \times \mathbf{T}^{\text{wt}}} J(\mathrm{d}T, \mathrm{d}S) \big(f(S) - f(T) \big) \big(g(S) - g(T) \big),$$

for f, g in the domain

$$\mathcal{D}^*(\mathcal{E}) := \{ f \in L^2(\mathbf{T}^{\text{wt}}, \mathbf{P}) : f \text{ is measurable, and } \mathcal{E}(f, f) < \infty \},$$

(here as usual, $L^2(\mathbf{T}^{\text{wt}}, \mathbf{P})$ is equipped with the inner product $(f, g)_{\mathbf{P}} := \int \mathbf{P}(\mathrm{d}x) f(x) g(x)$). By the argument in Example 1.2.1 in [72] and Lemma 9.5, $(\mathcal{E}, \mathcal{D}^*(\mathcal{E}))$ is well-defined, symmetric and Markovian.

Lemma 9.6. *The form $(\mathcal{E}, \mathcal{D}^*(\mathcal{E}))$ is closed. That is, if $(f_n)_{n \in \mathbb{N}}$ be a sequence in $\mathcal{D}^*(\mathcal{E})$ such that*

$$\lim_{m,n \to \infty} \big(\mathcal{E}(f_n - f_m, f_n - f_m) + (f_n - f_m, f_n - f_m)_{\mathbf{P}} \big) = 0,$$

then there exists $f \in \mathcal{D}^*(\mathcal{E})$ such that

$$\lim_{n \to \infty} \big(\mathcal{E}(f_n - f, f_n - f) + (f_n - f, f_n - f)_{\mathbf{P}} \big) = 0.$$

Proof. Let $(f_n)_{n \in \mathbb{N}}$ be a sequence such that $\lim_{m,n \to \infty} \mathcal{E}(f_n - f_m, f_n - f_m) + (f_n - f_m, f_n - f_m)_{\mathbf{P}} = 0$ (that is, $(f_n)_{n \in \mathbb{N}}$ is Cauchy with respect to $\mathcal{E}(\cdot, \cdot) + (\cdot, \cdot)_{\mathbf{P}}$). There exists a subsequence $(n_k)_{k \in \mathbb{N}}$ and $f \in L_2(\mathbf{T}^{\text{wt}}, \mathbf{P})$ such that $\lim_{k \to \infty} f_{n_k} = f$, $\mathbf{P}$-a.s, and $\lim_{k \to \infty} (f_{n_k} - f, f_{n_k} - f)_{\mathbf{P}} = 0$. By Fatou's Lemma,

$$\int J(\mathrm{d}T, \mathrm{d}S)\big((f(S) - f(T))\big)^2 \leqslant \liminf_{k \to \infty} \mathcal{E}(f_{n_k}, f_{n_k}) < \infty,$$

and so $f \in \mathcal{D}^*(\mathcal{E})$. Similarly,

$$\mathcal{E}(f_n - f, f_n - f)$$
$$= \int J(\mathrm{d}T, \mathrm{d}S) \lim_{k \to \infty} \big((f_n - f_{n_k})(S) - (f_n - f_{n_k})(T)\big)^2$$
$$\leqslant \liminf_{k \to \infty} \mathcal{E}(f_n - f_{n_k}, f_n - f_{n_k}) \to 0$$

as $n \to \infty$. Thus, $\{f_n\}_{n \in \mathbb{N}}$ has a subsequence that converges to f with respect to $\mathcal{E}(\cdot, \cdot) + (\cdot, \cdot)_\mathbf{P}$, but, by the Cauchy property, this implies that $\{f_n\}_{n \in \mathbb{N}}$ itself converges to f. $\square$

Let $\mathcal{L}$ denote the collection of functions $f : \mathbf{T}^{\mathrm{wt}} \to \mathbb{R}$ such that

$$\sup_{T \in \mathbf{T}^{\mathrm{wt}}} |f(T)| < \infty \qquad (9.11)$$

and

$$\sup_{S, T \in \mathbf{T}^{\mathrm{wt}}, \, S \neq T} \frac{|f(S) - f(T)|}{\Delta_{\mathrm{GH}^{\mathrm{wt}}}(S, T)} < \infty. \qquad (9.12)$$

Note that $\mathcal{L}$ consists of continuous functions and contains the constants. It follows from (4.20) that $\mathcal{L}$ is both a vector lattice and an algebra. By Lemma 9.7 below, $\mathcal{L} \subseteq \mathcal{D}^*(\mathcal{E})$. Therefore, the closure of $(\mathcal{E}, \mathcal{L})$ is a Dirichlet form that we will denote by $(\mathcal{E}, \mathcal{D}(\mathcal{E}))$.

Lemma 9.7. *Suppose that* $\{f_n\}_{n \in \mathbb{N}}$ *is a sequence of functions from* $\mathbf{T}^{\mathrm{wt}}$ *into* $\mathbb{R}$ *such that*

$$\sup_{n \in \mathbb{N}} \sup_{T \in \mathbf{T}^{\mathrm{wt}}} |f_n(T)| < \infty,$$

$$\sup_{n \in \mathbb{N}} \sup_{S, T \in \mathbf{T}^{\mathrm{wt}}, \, S \neq T} \frac{|f_n(S) - f_n(T)|}{\Delta_{\mathrm{GH}^{\mathrm{wt}}}(S, T)} < \infty,$$

and

$$\lim_{n \to \infty} f_n = f, \qquad \mathbf{P}\text{-}a.s.$$

for some $f : \mathbf{T}^{\mathrm{wt}} \to \mathbb{R}$. *Then* $\{f_n\}_{n \in \mathbb{N}} \subset \mathcal{D}^*(\mathcal{E})$, $f \in \mathcal{D}^*(\mathcal{E})$, *and*

$$\lim_{n \to \infty} \big(\mathcal{E}(f_n - f, f_n - f) + (f_n - f, f_n - f)_\mathbf{P}\big) = 0.$$

Proof. By the definition of the measure J, see (9.4), and the symmetry of J (Lemma 9.5(i)), we have that $f_n(x) - f_n(y) \to f(x) - f(y)$ for J-almost every pair (x, y). The result then follows from part (iii) of Lemma 9.5 and the dominated convergence theorem. $\square$

Before showing that $(\mathcal{E}, \mathcal{D}(\mathcal{E}))$ is the Dirichlet form of a nice Markov process, we remark that $\mathcal{L}$, and thus also $\mathcal{D}(\mathcal{E})$, is quite a rich class of functions. We show in the proof of Theorem 9.8 below that $\mathcal{L}$ separates points of $\mathbf{T}^{\mathrm{wt}}$. Hence, if $\mathbf{K}$ is any compact subset of $\mathbf{T}^{\mathrm{wt}}$, then, by the Arzela-Ascoli theorem, the set of restrictions of functions in $\mathcal{L}$ to $\mathbf{K}$ is uniformly dense in the space of real-valued continuous functions on $\mathbf{K}$.

The following theorem states that there is a well-defined Markov process with the dynamics we would expect for a limit of the subtree prune and regraft chains.

Theorem 9.8. *There exists a recurrent $\mathbf{P}$-symmetric Hunt process $X = (X_t, \mathbb{P}^T)$ on $\mathbf{T}^{\mathrm{wt}}$ whose Dirichlet form is $(\mathcal{E}, \mathcal{D}(\mathcal{E}))$.*

Proof. We will check the conditions of Theorem A.8 to establish the existence of X.

Because $\mathbf{T}^{\mathrm{wt}}$ is complete and separable (recall Theorem 4.44) there is a sequence $\mathbf{H}_1 \subseteq \mathbf{H}_2 \subseteq \ldots$ of compact subsets of $\mathbf{T}^{\mathrm{wt}}$ such that $\mathbf{P}(\bigcup_{k \in \mathbb{N}} \mathbf{H}_k) = 1$. Given $\alpha, \beta > 0$, write $\mathcal{L}_{\alpha,\beta}$ for the subset of $\mathcal{L}$ consisting of functions f such that

$$\sup_{T \in \mathbf{T}^{\mathrm{wt}}} |f(T)| \leqslant \alpha$$

and

$$\sup_{S,T \in \mathbf{T}^{\mathrm{wt}}, S \neq T} \frac{|f(S) - f(T)|}{\Delta_{\mathrm{GH}^{\mathrm{wt}}}(S,T)} \leqslant \beta.$$

By the separability of the continuous real-valued functions on each $\mathbf{H}_k$ with respect to the supremum norm, it follows that for each $k \in \mathbb{N}$ there is a countable set $L_{\alpha,\beta,k} \subseteq \mathcal{L}_{\alpha,\beta}$ such that for every $f \in \mathcal{L}_{\alpha,\beta}$

$$\inf_{g \in L_{\alpha,\beta,k}} \sup_{T \in H_k} |f(T) - g(T)| = 0.$$

Set $L_{\alpha,\beta} := \bigcup_{k \in \mathbb{N}} L_{\alpha,\beta,k}$. Then for any $f \in \mathcal{L}_{\alpha,\beta}$ there exists a sequence $\{f_n\}_{n \in \mathbb{N}}$ in $L_{\alpha,\beta}$ such that $\lim_{n \to \infty} f_n = f$ pointwise on $\bigcup_{k \in \mathbb{N}} \mathbf{H}_k$, and, *a fortiori*, $\mathbf{P}$-almost surely. By Lemma 9.7, the countable set $\bigcup_{m \in \mathbb{N}} L_{m,m}$ is dense in $\mathcal{L}$ and, *a fortiori*, in $\mathcal{D}(\mathcal{E})$, with respect to $\mathcal{E}(\cdot, \cdot) + (\cdot, \cdot)_{\mathbf{P}}$.

Now fix a countable dense subset $\mathbf{S} \subset \mathbf{T}^{\mathrm{wt}}$. Let M denote the countable set of functions of the form

$$T \mapsto p + q(\Delta_{\mathrm{GH}^{\mathrm{wt}}}(S,T) \wedge r)$$

for some $S \in \mathbf{S}$ and $p, q, r \in \mathbb{Q}$. Note that $M \subseteq \mathcal{L}$, that M separates the points of $\mathbf{T}^{\mathrm{wt}}$, and, for any $T \in \mathbf{T}^{\mathrm{wt}}$, that there is certainly a function $f \in M$ with $f(T) \neq 0$.

Consequently, if $\mathcal{C}$ is the algebra generated by the countable set $M \cup \bigcup_{m \in \mathbb{N}} L_{m,m}$, then it is certainly the case that $\mathcal{C}$ is dense in $\mathcal{D}(\mathcal{E})$ with respect $\mathcal{E}(\cdot, \cdot) + (\cdot, \cdot)_{\mathbf{P}}$, that $\mathcal{C}$ separates the points of $\mathbf{T}^{\mathrm{wt}}$, and, for any $T \in \mathbf{T}^{\mathrm{wt}}$, that there is a function $f \in \mathcal{C}$ with $f(T) \neq 0$.

All that remains in verifying the conditions of Theorem A.8 is to check the tightness condition that there exist compact subsets $\mathbf{K}_1 \subseteq \mathbf{K}_2 \subseteq \dots$ of $\mathbf{T}^{\mathrm{wt}}$ such that $\lim_{n\to\infty} \mathrm{Cap}(\mathbf{T}^{\mathrm{wt}} \backslash \mathbf{K}_n) = 0$ where Cap is the capacity associated with the Dirichlet form This convergence, however, is the content of Lemma 9.11 below.

Finally, because constants belongs to $\mathcal{D}(\mathcal{E})$, it follows from Theorem 1.6.3 in [72] that X is recurrent. $\qquad\qquad\square$

The following results were needed in the proof of Theorem 9.8

Lemma 9.9. *For $\varepsilon, a, \delta > 0$, put $\mathbf{V}_{\varepsilon,a} := \{T \in \mathbf{T} : \mu^T(R_\varepsilon(T)) > a\}$ and, as usual, $\mathbf{V}_{\varepsilon,a}^\delta := \{T \in \mathbf{T} : d_{\mathrm{GH}}(T, \mathbf{V}_{\varepsilon,a}) < \delta\}$. Then, for fixed $\varepsilon > 3\delta$,*

$$\bigcap_{a>0} \mathbf{V}_{\varepsilon,a}^\delta = \varnothing.$$

Proof. Fix $S \in \mathbf{T}$. If $S \in \mathbf{V}_{\varepsilon,a}^\delta$, then there exists $T \in \mathbf{V}_{\varepsilon,a}$ such that $d_{\mathrm{GH}}(S,T) < \delta$. Observe that $R_\varepsilon(T)$ is not the trivial tree consisting of a single point because it has total length greater than a. Write $\{y_1, \dots, y_n\}$ for the leaves of $R_\varepsilon(T)$. Note that $T \backslash R_\varepsilon(T)^\circ$ is the union of n subtrees of diameter ε. The closure of each subtree contains a unique y_i. Choose z_i in the subtree whose closure contains y_i such that $d_T(y_i, z_i) = \varepsilon$.

Let $\mathfrak{R}$ be a correspondence between S and T with $\mathrm{dis}(\mathfrak{R}) < 2\delta$. Pick $x_1, \dots, x_n \in S$ such that $(x_i, z_i) \in \mathfrak{R}$. Hence, $|d_S(x_i, x_j) - d_T(z_i, z_j)| < 2\delta$ for all i, j.

The distance in $R_\varepsilon(T)$ from the point y_k to the segment $[y_i, y_j]$ is

$$\frac{1}{2}\big(d_S(y_k, y_i) + d_S(y_k, y_j) - d_S(y_i, y_j)\big).$$

Thus, the distance from y_k, $3 \leqslant k \leqslant n$, to the subtree spanned by $y_1, \dots, y_{k-1}$ is

$$\bigwedge_{1\leqslant i\leqslant j\leqslant k-1} \frac{1}{2}\big(d_T(y_k, y_i) + d_T(y_k, y_j) - d_T(y_i, y_j)\big).$$

Hence,

$$\mu^T(R_\varepsilon(T)) = d_T(y_1, y_2)$$

$$+ \sum_{k=3}^{n} \bigwedge_{1\leqslant i\leqslant j\leqslant k-1} \frac{1}{2}\big(d_T(y_k, y_i) + d_T(y_k, y_j) - d_T(y_i, y_j)\big).$$

Now the distance in S from the point x_k to the segment $[x_i, x_j]$ is

$$\frac{1}{2}\big(d_S(x_k, x_i) + d_S(x_k, x_j) - d_S(x_i, x_j)\big)$$

$$\geqslant \frac{1}{2}\big(d_T(z_k, z_i) + d_T(z_k, z_j) - d_T(z_i, z_j) - 3 \times 2\delta\big)$$

$$= \frac{1}{2}\big(d_T(y_k, y_i) + 2\varepsilon + d_T(y_k, y_j) + 2\varepsilon - d_T(y_i, y_j) - 2\varepsilon - 6\delta\big)$$

$$> 0$$

by the assumption that $\varepsilon > 3\delta$. In particular, $x_1, \ldots, x_n$ are leaves of the subtree spanned by $\{x_1, \ldots, x_n\}$, and $R_\gamma(S)$ has at least n leaves when $0 < \gamma < 2\varepsilon - 6\delta$. Fix such a γ.

Now

$$
\mu^S(R_\gamma(S))
$$
$$
\geqslant d_S(x_1, x_2) - 2\gamma
$$
$$
+ \sum_{k=3}^{n} \bigwedge_{1 \leqslant i \leqslant j \leqslant k-1} \left[\frac{1}{2} \left(d_S(x_k, x_i) + d_S(x_k, x_j) - d_S(x_i, x_j) \right) - \gamma \right]
$$
$$
\geqslant \mu^T(R_\varepsilon(T)) + (2\varepsilon - 2\delta - 2\gamma) + (n - 2)(\varepsilon - 3\delta - \gamma)
$$
$$
\geqslant a + (2\varepsilon - 2\delta - 2\gamma) + (n - 2)(\varepsilon - 3\delta - \gamma).
$$

Because $\mu^S(R_\gamma(S))$ is finite, it is apparent that S cannot belong to $\mathbf{V}^\delta_{\varepsilon,a}$ when a is sufficiently large.   □

Lemma 9.10. *For $\varepsilon, a, \delta > 0$, let $\mathbf{V}_{\varepsilon,a}$ be as in Lemma 9.9. Set $\mathbf{U}_{\varepsilon,a} := \{(T, \nu) \in \mathbf{T}^{\mathrm{wt}} : T \in \mathbf{V}_{\varepsilon,a}\}$. Then, for fixed ε,*

$$
\lim_{a \to \infty} \mathrm{Cap}(\mathbf{U}_{\varepsilon,a}) = 0.
$$

Proof. Choose $\delta > 0$ such that $\varepsilon > 3\delta$. Suppressing the dependence on ε and δ, define $u_a : \mathbf{T}^{\mathrm{wt}} \to [0,1]$ by

$$
u_a((T, \nu)) := \delta^{-1}(\delta - d_{\mathrm{GH}}(T, \mathbf{V}_{\varepsilon,a}))_+.
$$

Note that u_a takes the value 1 on the open set $\mathbf{U}_{\varepsilon,a}$, and so $\mathrm{Cap}(\mathbf{U}_{\varepsilon,a}) \leqslant \mathcal{E}(u_a, u_a) + (u_a, u_a)_{\mathbf{P}}$. Also, observe that

$$
|u_a((T', \nu')) - u_a((T'', \nu''))| \leqslant \delta^{-1} d_{\mathrm{GH}}(T', T'')
$$
$$
\leqslant \delta^{-1} \Delta_{\mathrm{GH}^{\mathrm{wt}}}((T', \nu'), (T'', \nu'')).
$$

It suffices, therefore, by part (iii) of Lemma 9.5 and the dominated convergence theorem to show for each pair $((T', \nu'), (T'', \nu'')) \in \mathbf{T}^{\mathrm{wt}} \times \mathbf{T}^{\mathrm{wt}}$ that $u_a((T', \nu')) - u_a((T'', \nu''))$ is 0 for a sufficiently large and for each $T \in \mathbf{T}^{\mathrm{wt}}$ that $u_a((T, \nu))$ is 0 for a sufficiently large. However, $u_a((T', \nu')) - u_a((T'', \nu'')) \neq 0$ implies that either T' or T'' belong to $\mathbf{V}^\delta_{\varepsilon,a}$, while $u_a((T, \nu)) \neq 0$ implies that T belongs to $\mathbf{V}^\delta_{\varepsilon,a}$. The result then follows from Lemma 9.9.   □

Lemma 9.11. *There is a sequence of compact sets $\mathbf{K}_1 \subseteq \mathbf{K}_2 \subseteq \ldots$ such that*

$$
\lim_{n \to \infty} \mathrm{Cap}(\mathbf{T}^{\mathrm{wt}} \backslash \mathbf{K}_n) = 0.
$$

Proof. By Lemma 9.10, for $n = 1, 2, \ldots$ we can choose a_n so that

$$
\mathrm{Cap}(\mathbf{U}_{2^{-n}, a_n}) \leqslant 2^{-n}.
$$

Set
$$\mathbf{F}_n := \mathbf{T}^{\mathrm{wt}} \backslash \mathbf{U}_{2^{-n}, a_n} = \{(T, \nu) \in \mathbf{T}^{\mathrm{wt}} : \mu^T(R_{2^{-n}}(T)) \leqslant a_n\}$$
and
$$\mathbf{K}_n := \bigcap_{m \geqslant n} \mathbf{F}_n.$$

By Proposition 4.43 and the analogue of Corollary 4.38 for unrooted trees, each set $\mathbf{K}_n$ is compact. By construction,

$$\begin{aligned} \mathrm{Cap}(\mathbf{T}^{\mathrm{wt}} \backslash \mathbf{K}_n) &= \mathrm{Cap}\left(\bigcup_{m \geqslant n} \mathbf{U}_{2^{-m}, a_m} \right) \\ &\leqslant \sum_{m \geqslant n} \mathrm{Cap}(\mathbf{U}_{2^{-m}, a_m}) \leqslant \sum_{m \geqslant n} 2^{-m} = 2^{-(n-1)}. \end{aligned}$$

$\square$

A

Summary of Dirichlet Form Theory

Our treatment in this appendix follows that of the standard reference [72] –
see also, [106, 3].

A.1 Non-Negative Definite Symmetric Bilinear Forms

Let H be a real Hilbert space with inner product $(\cdot, \cdot)$. We say $\mathcal{E}$ is a *non-
negative definite symmetric bilinear form* on H with domain $\mathcal{D}(\mathcal{E})$ if

- $\mathcal{D}(\mathcal{E})$ is a dense linear subspace of H,
- $\mathcal{E} : \mathcal{D}(\mathcal{E}) \times \mathcal{D}(\mathcal{E}) \to \mathbb{R}$,
- $\mathcal{E}(u, v) = \mathcal{E}(v, u)$ for $u, v \in \mathcal{D}(\mathcal{E})$,
- $\mathcal{E}(au + bv, w) = a\mathcal{E}(u, w) + b\mathcal{E}(v, w)$ for $u, v, w \in \mathcal{D}(\mathcal{E})$ and $a, b \in \mathbb{R}$,
- $\mathcal{E}(u, u) \geqslant 0$ for $u \in \mathcal{D}(\mathcal{E})$.

Given a non-negative definite symmetric bilinear form $\mathcal{E}$ on H and $\alpha > 0$,
define another non-negative definite symmetric bilinear form $\mathcal{E}_\alpha$ on H with
domain $\mathcal{D}(\mathcal{E}_\alpha) := \mathcal{D}(\mathcal{E})$ by

$$\mathcal{E}_\alpha(u, v) := \mathcal{E}(u, v) + \alpha(u, v), \quad u, v \in \mathcal{D}(\mathcal{E}).$$

Note that the space $\mathcal{D}(\mathcal{E})$ is a pre-Hilbert space with inner product $\mathcal{E}_\alpha$, and
$\mathcal{E}_\alpha$ and $\mathcal{E}_\beta$ determine equivalent metrics on $\mathcal{D}(\mathcal{E})$ for different $\alpha, \beta > 0$.

If $\mathcal{D}(\mathcal{E})$ is complete with respect to this metric, then $\mathcal{E}$ is said to be *closed*.
In this case, $\mathcal{D}(\mathcal{E})$ is then a real Hilbert space with inner product $\mathcal{E}_\alpha$ for each
$\alpha > 0$.

A.2 Dirichlet Forms

Now consider a σ-finite measure space $(X, \mathcal{B}, m)$ and take H to be the Hilbert
space $L^2(X, m)$ with the usual inner product

$$(u, v) := \int_X u(x)v(x)\, m(dx), \quad u, v \in L^2(X, m).$$

Call a non-negative definite symmetric bilinear form $\mathcal{E}$ on $L^2(X, m)$ *Markovian* if for each $\varepsilon > 0$, there exists a real function $\phi_\varepsilon : \mathbb{R} \to \mathbb{R}$, such that

$$\phi_\varepsilon(t) = t, \quad t \in [0, 1],$$
$$-\varepsilon \leqslant \phi_\varepsilon(t) \leqslant 1 + \varepsilon, \quad t \in \mathbb{R},$$
$$0 \leqslant \phi_\varepsilon(t) - \phi_\varepsilon(s) \leqslant t - s, \quad -\infty < s < t < \infty,$$

and when u belongs to $\mathcal{D}(\mathcal{E})$, $\phi_\varepsilon \circ u$ also belongs to $\mathcal{D}(\mathcal{E})$ with

$$\mathcal{E}(\phi_\varepsilon \circ u, \phi_\varepsilon \circ u) \leqslant \mathcal{E}(u, u).$$

A *Dirichlet form* is a non-negative definite symmetric bilinear form on $L^2(X, m)$ that is Markovian and closed.

A non-negative definite symmetric bilinear form $\mathcal{E}$ on $L^2(X, m)$ is certainly Markovian if whenever u belongs to $\mathcal{D}(\mathcal{E})$, then $v = (0 \vee u) \wedge 1$ also belongs to $\mathcal{D}(\mathcal{E})$ and $\mathcal{E}(v, v) \leqslant \mathcal{E}(u, u)$. In this case say that the *unit contraction* acts on $\mathcal{E}$. It turns out the if the form is closed, then the form is Markovian if and only if the unit contraction acts on it.

Similarly, say that a function v is called a *normal contraction* of a function u if

$$|v(x) - v(y)| \leqslant |u(x) - u(y)|, \quad x, y \in X,$$
$$|v(x)| \leqslant |u(x)|, \quad x \in X,$$

and say that $v \in L^2(X, m)$ a normal contraction of $u \in L^2(X, m)$ if some Borel version of v is a normal contraction of some Borel version of u. Say that normal contractions act on $\mathcal{E}$ if whenever v is a normal contraction of $u \in \mathcal{D}(\mathcal{E})$, then $v \in \mathcal{D}(\mathcal{E})$ and $\mathcal{E}(v, v) \leqslant \mathcal{E}(u, u)$. It also turns out that if the form is closed, then the form is Markovian if and only if the unit contraction acts on it.

Example A.1. Let $X \subseteq \mathbb{R}$ be an open subinterval and suppose that m is a Radon measures on X with support all of X. Define a non-negative definite symmetric bilinear form by

$$\mathcal{E}(u, v) := \frac{1}{2} \int_X \frac{du(x)}{dx} \frac{dv(x)}{dx}\, dx$$

on the domain

$$\mathcal{D}(\mathcal{E}) := \{u \in L^2(X, m) : u \text{ is absolutely continuous and } \mathcal{E}(u, u) < \infty\}.$$

We claim that $\mathcal{E}$ is a Dirichlet form on $L^2(X, m)$.

It is easy to check that the unit contraction acts on $\mathcal{E}$. To show the form is closed, take any $\mathcal{E}_1$-Cauchy sequence $\{u_\ell\}$. Then $\{du_\ell/dx\}$ converges to some $f \in L^2(X, dx)$ in $L^2(X, dx)$. Also, $\{u_\ell\}$ converges to some $u \in L^2(X, m)$ in $L^2(X, m)$. From this and the inequality

$$|u(a) - u(b)|^2 \leqslant 2|a - b|\mathcal{E}(u, u), \quad a, b \in X,$$

we conclude that there is a subsequence $\{\ell_k\}$ such that u_{ℓ_k} converges to a continuous function $\tilde{u}$ uniformly on each bounded closed subinterval of X. Obviously $\tilde{u} = u$ m-a.e. For all infinitely differentiable compactly supported functions ϕ on X, an integration by parts shows that

$$\int_X f(x)\phi(x)\, dx = \lim_{\ell_k \to \infty} \int_X \frac{du_{\ell_k}(x)}{dx} \phi(x)\, dx$$

$$= -\lim_{\ell_k \to \infty} \int_X u_{\ell_k}(x)\phi'(x)\, dx = -\int_X \tilde{u}(x)\phi'(x)\, dx.$$

This implies that $\tilde{u}$ is absolutely continuous and $d\tilde{u}/dx = f$. Hence, $\tilde{u} \in \mathcal{D}(\mathcal{E})$ and $\{u_\ell\}$ is $\mathcal{E}_1$-convergent to $\tilde{u}$.

Example A.2. Consider a locally compact metric space (X, ρ) equipped with a Radon measure m. Suppose that we are given a kernel j on $X \times \mathcal{B}(X)$ satisfying the following conditions.

- For any $\varepsilon > 0$, $j(x, X \backslash B_\varepsilon(x))$ is, as a function of $x \in X$, locally integrable with respect to m. Here, as usual, $B_\varepsilon(x)$ is the ball around x of radius ε.
- $\int_X u(x)\,(jv)(x)\, m(dx) = \int_X (ju)(x)\, v(x)\, m(dx)$ for all $u, v \in p\mathcal{B}(X)$.

Then, j determines a symmetric Radon measure J on $X \times X \backslash \Delta$, where Δ is the diagonal, by

$$\int_{X \times X \backslash \Delta} f(x, y)\, J(dx, dy) := \int_X \left\{ \int_X f(x, y)\, j(x, dy) \right\} m(dx).$$

Put

$$\mathcal{E}(u, v) := \int_{X \times X \backslash \Delta} (u(x) - u(y))(v(x) - v(y))\, J(dx, dy)$$

on the domain

$$\mathcal{D}(\mathcal{E}) := \{u \in L^2(X, m) : \mathcal{E}(u, u) < \infty\}.$$

We claim that $\mathcal{E}$ is a Dirichlet form on $L^2(X, m)$ provided that $\mathcal{D}(\mathcal{E})$ is dense in $L^2(X, m)$.

It is clear that $\mathcal{E}$ is non-negative definite, symmetric, and bilinear. We next show that for a Borel function u that $u = 0$ m-a.e. implies that $\mathcal{E}(u, u) = 0$. Suppose that $u = 0$ m-a.e. Put $\Gamma_{K,\varepsilon} = \{(x, y) \in K \times K : \rho(x, y) > \varepsilon\}$ for $\varepsilon > 0$ and K compact. Then

$$\int_{\Gamma_{K,\varepsilon}} (u(x) - u(y))^2 \, J(dx, dy) \leqslant 2 \int_{\Gamma_{K,\varepsilon}} (u(x)^2 + u(y)^2) \, J(dx, dy)$$

$$= 4 \int_{\Gamma_{K,\varepsilon}} u(x)^2 J(dx, dy) \leqslant 4 \int_K u(x)^2 j(x, X \backslash B_\varepsilon(x)) \, m(dx) = 0.$$

Letting $\varepsilon \downarrow 0$ and $K \uparrow X$ gives $\mathcal{E}(u, u) = 0$.

It is clear that every normal contraction operates on the form and so the form is Markovian. To prove that the form is closed, consider a sequence $\{u_\ell\}$ in $\mathcal{D}(\mathcal{E})$ such that $\lim_{\ell, m \to \infty} \mathcal{E}_1(u_\ell - u_m, u_\ell - u_m) \to 0$. Since $\{u_\ell\}$ converges in $L^2(X, m)$, there is a subsequence $\{\ell_k\}$ and a set $N \in \mathcal{B}(X)$ with $m(N) = 0$ such that $\{u_{\ell_k}(x)\}$ converges on $X \backslash N$. Put $\tilde{u}_{\ell_k}(x) = u_{l_k}(x)$ on $X \backslash N$ and $\tilde{u}_{\ell_k}(x) = 0$ on N. Then $\tilde{u}_{\ell_k}(x)$ has a limit $u(x)$ everywhere and u_ℓ converges to u in $L^2(X, m)$. Moreover,

$$\mathcal{E}(u - u_m, u - u_m)$$

$$= \int_{X \times X \backslash \Delta} \lim_{\ell_k \to \infty} \{(u_{\ell_k}(x) - u_{\ell_k}(y)) - (u_m(x) - u_m(y))\}^2 \, J(dx, dy)$$

$$\leqslant \liminf_{l_k \to \infty} \mathcal{E}(u_{l_k} - u_m, u_{l_k} - u_m).$$

The last term can be made arbitrarily small for sufficiently large m. Thus, u_m is $\mathcal{E}_1$-convergent to $u \in \mathcal{D}(\mathcal{E})$, as required.

A.3 Semigroups and Resolvents

Suppose again that we have a real Hilbert space H with inner product $(\cdot, \cdot)$. Consider a family $\{T_t\}_{t > 0}$ of linear operators on H satisfying the following conditions:

- each T_t is a self-adjoint operator with domain H,
- $T_s T_t = T_{s+t}$, $s, t > 0$ (that is, $\{T_t\}_{t > 0}$ is a *semigroup*),
- $(T_t u, T_t u) \leqslant (u, u)$, $t > 0$, $u \in H$ (that is, each T_t is a *contraction*).

We say that $\{T_t\}_{t > 0}$ is *strongly continuous* if, in addition,

- $\lim_{t \downarrow 0}(T_t u - u, T_t u - u) = 0$ for all $u \in H$.

A *resolvent* on H is a family $\{G_\alpha\}_{\alpha > 0}$ of linear operators on H satisfying the following conditions:

- G_α is a self-adjoint operator with domain H,
- $G_\alpha - G_\beta + (\alpha - \beta)G_\alpha G_\beta = 0$ (the resolvent equation),
- each operator αG_α is a contraction.

The resolvent is said to be *strongly continuous* if, in addition,

- $\lim_{\alpha \to \infty}(\alpha G_\alpha u - u, \alpha G_\alpha u - u) = 0$ for all $u \in H$.

Example A.3. Given a strongly continuous semigroup $\{T_t\}_{t>0}$ on H, the family of operators

$$G_\alpha u := \int_0^\infty e^{-\alpha t} T_t u \, dt$$

is a strongly continuous resolvent on H called the resolvent of the given semigroup. The semigroup may be recovered from the resolvent via the *Yosida approximation*

$$T_t u = \lim_{\beta \to \infty} e^{-t\beta} \sum_{n=0}^\infty \frac{(t\beta)^n}{n!} (\beta G_\beta)^n u, \quad u \in H.$$

A.4 Generators

The *generator A* of a strongly continuous semigroup $\{T_t\}_{t>0}$ on H is defined by

$$Au := \lim_{t \downarrow 0} \frac{T_t u - u}{t}$$

on the domain $\mathcal{D}(A)$ consisting of those $u \in H$ such that the limit exists.

Suppose that $\{G_\alpha\}_{\alpha>0}$ is a strongly continuous resolvent on H. Note that if $G_\alpha u = 0$, then, by the resolvent equation, $G_\beta u = 0$ for all $\beta > 0$, and, by strong continuity, $u = \lim_{\beta \to \infty} \beta G_\beta u = 0$. Thus, the operator G_α is invertible and we can set

$$Au := \alpha u - G_\alpha^{-1} u$$

on the domain $\mathcal{D}(A) := G_\alpha(H)$. This operator A is easily seen to be independent of $\alpha > 0$ and is called the *generator* of the resolvent. $\{G_\alpha\}_{\alpha>0}$.

Lemma A.4. *The generator of a strongly continuous semigroup on H coincides with the generator of its resolvent, and the generator is a non-positive definite self-adjoint operator.*

A.5 Spectral Theory

A self-adjoint operator S on H with domain H satisfying $S^2 = S$ is called a *projection*. A family $\{E_\lambda\}_{\lambda \in \mathbb{R}}$ of projection operators on H is called a *spectral family* if

$$E_\lambda E_\mu = E_\lambda, \quad \lambda \leqslant \mu,$$

$$\lim_{\lambda' \downarrow \lambda} E_{\lambda'} u = E_\lambda u, \quad u \in H,$$

$$\lim_{\lambda \to -\infty} E_\lambda u = 0, \quad u \in H,$$

$$\lim_{\lambda \to \infty} E_\lambda u = u, \quad u \in H.$$

Note that $0 \leqslant (E_\lambda u, u) \uparrow (u, u)$ as $\lambda \uparrow \infty$, for $u \in H$, and, by polarization, $\lambda \mapsto (E_\lambda u, v)$ is a function of bounded variation for $u, v \in H$.

Suppose we are given a spectral family $\{E_\lambda\}_{\lambda \in \mathbb{R}}$ on H and a continuous function $\phi(\lambda)$ on $\mathbb{R}$. We can then define a self-adjoint operator A on H, denoted by $\int_{-\infty}^{\infty} \phi(\lambda) \, dE_\lambda$, by requiring that

$$(Au, v) = \int_{-\infty}^{\infty} \phi(\lambda) \, d(E_\lambda u, v), \quad \forall v \in H,$$

where the domain of A is $\mathcal{D}(A) := \{u \in H : \int_{-\infty}^{\infty} \phi(\lambda) \, d(E_\lambda u, u) < \infty\}$.

Conversely, given a self-adjoint operator A on H, there exists a unique spectral family $\{E_\lambda\}_{\lambda \in \mathbb{R}}$ such that $A = \int_{-\infty}^{\infty} \lambda \, dE_\lambda$. This is called the spectral representation of A. If A is non-negative definite, then the corresponding spectral family satisfies $E_\lambda = 0$ for $\lambda < 0$.

Let $-A$ be a non-negative definite self-adjoint operator on H and let $-A = \int_0^{\infty} \lambda \, dE_\lambda$ be its spectral representation. For any non-negative continuous function ϕ on $\mathbb{R}_+$, we define the self-adjoint operator $\phi(-A)$ by $\phi(-A) := \int_0^{\infty} \phi(\lambda) \, dE_\lambda$. Note that $\phi(-A)$ is again non-negative definite.

A.6 Dirichlet Form, Generator, Semigroup, Resolvent Correspondence

Lemma A.5. *Let $-A$ be a non-negative definite self-adjoint operator on H. The family $\{T_t\}_{t>0} := \{\exp(tA)\}_{t>0}$ is a strongly continuous semigroup, and the family $\{G_\alpha\}_{\alpha>0} := \{(\alpha - A)^{-1}\}_{\alpha>0}$ is a strongly continuous resolvent. The generator of $\{T_t\}_{t>0}$ is A and $\{T_t\}_{t>0}$ is the unique strongly continuous semigroup with generator A. A similar statement holds for the resolvent.*

Theorem A.6. *There is a bijective correspondence between the family of closed non-negative definite symmetric bilinear forms $\mathcal{E}$ on H and the family of non-positive definite self-adjoint operators A on H. The correspondence is given by*

$$\mathcal{D}(\mathcal{E}) = \mathcal{D}(\sqrt{-A})$$

and

$$\mathcal{E}(u, v) = (\sqrt{-A}u, \sqrt{-A}v).$$

Consider a σ-finite measure space $(X, \mathcal{B}, m)$. A linear operator S on $L^2(X, m)$ with domain $L^2(X, m)$ is *Markovian* if $0 \leqslant Su \leqslant 1$ m-a.e. whenever $u \in L^2(X, m)$ and $0 \leqslant u \leqslant 1$ m-a.e.

Theorem A.7. *Let $\mathcal{E}$ be a closed non-negative definite symmetric bilinear form on $L^2(X, m)$. Write $\{T_t\}_{t>0}$ and $\{G_\alpha\}_{\alpha>0}$ for the corresponding strongly continuous semigroup and the strongly continuous resolvent on $L^2(X, m)$. The following five conditions are equivalent.*

(a) T_t is Markovian for each $t > 0$.
(b) αG_α is Markovian for each $\alpha > 0$.
(c) $\mathcal{E}$ is Markovian.
(d) The unit contraction operates on $\mathcal{E}$.
(e) Normal contractions operate on $\mathcal{E}$.

A.7 Capacities

Suppose that X is a Lusin space and m is a Radon measure. There is a set function associated with a Dirichlet form $(\mathcal{E}, \mathcal{D}(\mathcal{E}))$ on $L^2(X, m)$ called the (1)-capacity and denoted by Cap. If $U \subseteq X$ is open, then

$$\mathrm{Cap}(U) := \inf \left\{ \mathcal{E}_1(f, f) \, : \, f \in \mathcal{D}(\mathcal{E}), \, f(x) \geq 1, \, m - \text{a.e. } x \in U \right\}.$$

More generally, if $V \subseteq X$ is an arbitrary subset, then

$$\mathrm{Cap}(V) := \inf \left\{ \mathrm{Cap}(U) : V \subseteq U, \, U \text{ is open} \right\}.$$

The set function Cap is a Choquet capacity.

We say that some property holds quasi-everywhere or, equivalently, for quasi-every $x \in X$, if the set $x \in X$ where the property fails to hold has capacity 0. We abbreviate this by saying that the property holds q.e. or for q.e. every $x \in X$.

A.8 Dirichlet Forms and Hunt Processes

A *Hunt process* is a strong Markov process

$$\mathbf{X} = (\Omega, \mathcal{F}, \{\mathcal{F}_t\}_{t \geq 0}, \{\mathbb{P}^x\}_{x \in E}, \{X_t\}_{t \geq 0})$$

on a Lusin state space E that has right-continuous, left-limited sample paths and is also quasi-left-continuous. Write $\{P_t\}_{t \geq 0}$ for the transition semigroup of $\mathbf{X}$. That is, $P_t f(x) = \mathbb{P}^x[f(X_t)]$ for $f \in b\mathcal{B}(E)$. If μ is a Radon measure on $(E, \mathcal{B}(E))$, we say that $\mathbf{X}$ is μ-*symmetric* if $\int_E f(x) \, P_t g(x) \, \mu(dx) = \int_E P_t f(x) \, g(x) \, \mu(dx)$ for all $f, g \in b\mathcal{B}(E)$. Intuitively, if the process $\mathbf{X}$ is started according to the initial "distribution" μ, then reversing the direction of time leaves finite-dimensional distributions unchanged.

Theorem A.8. *Let $(\mathcal{E}, \mathcal{D}(\mathcal{E}))$ be a Dirichlet form on $L^2(E, \mu)$, where E is Lusin and μ is Radon. Write $\{T_t\}_{t > 0}$ for the associated strongly continuous contraction semigroup of Markovian operators. Suppose that there exists a collection $\mathcal{C} \subseteq L^2(E, \mu)$ and a sequence of compact sets $K_1 \subseteq K_2 \subseteq \ldots$ such that:*

(a) $\mathcal{C}$ *is a countably generated subalgebra of $\mathcal{D}(\mathcal{E}) \cap bC(E)$,*

(b) C is $\mathcal{E}_1$-dense in $\mathcal{D}(\mathcal{E})$,

(c) C separates points of E and, for any $x \in E$, there is an $f \in C$ such that $f(x) \neq 0$,

(d) $\lim_{n \to \infty} \mathrm{Cap}(E \backslash K_n) = 0$.

Then there is a μ-symmetric Hunt process $\mathbf{X}$ on E with transition semigroup $\{P_t\}_{t \geqslant 0}$ such that $P_t f(x) = T_t f(x)$ for $f \in b\mathcal{B}(E) \cap L^2(E, \mu)$.

Remark A.9. The theory in [72] for symmetric Hunt processes associated with Dirichlet forms is developed under the hypothesis that the state space is locally compact. However, the embedding results outlined in Section 7.3 of [72], shows that the results developed under the hypothesis of local compactness still holds if the state space is Lusin and the hypotheses of Theorem A.8 hold.

Lemma A.10. *Suppose that $\mathbf{X}$ is the μ-symmetric Hunt process constructed from a Dirichlet form $(\mathcal{E}, \mathcal{D}(\mathcal{E}))$ satisfying the conditions of Theorem A.8 and $B \in \mathcal{B}(E)$. Then $\mathbb{P}^x\{\exists t > 0 : X_t \in B\} = 0$ for μ-a.e. $x \in E$ if and only if $\mathrm{Cap}(B) = 0$.*

B

Some Fractal Notions

This appendix is devoted to recalling briefly some definitions about various ways of assigning sizes and dimensions to metric spaces and then applying this theory to the ultrametric completions of $\mathbb{N}$ obtained in Example 3.41 from the $\mathbb{R}$-tree associated with a non-increasing family of partitions of $\mathbb{N}$.

B.1 Hausdorff and Packing Dimensions

Let $(\mathcal{X}, \rho)$ be a compact metric space. Given a set $A \subseteq \mathcal{X}$ and $\epsilon > 0$, a countable collection of balls $\{B_i\}$ is said to be an ϵ-covering of A if $A \subseteq \bigcup_i B_i$ and each ball has diameter at most ϵ. Note that if $\epsilon' < \epsilon''$, then an ϵ'-covering of A is also an ϵ''-covering. For $\alpha > 0$, the α-dimensional Hausdorff measure on $\mathcal{X}$ is the Borel measure that assigns mass

$$\mathcal{H}^\alpha(A) := \sup_{\epsilon > 0} \inf \left\{ \sum_i \operatorname{diam}(B_i)^\alpha \ : \ \{B_i\} \text{ is an } \epsilon\text{-covering of } A \right\}$$

to a Borel set A. The *Hausdorff dimension* of A is the infimum of those α such that the corresponding α-dimensional Hausdorff measure is zero.

A countable collection of balls $\{B_i\}$ is said to be an ϵ-*packing* of a set $A \subseteq \mathcal{X}$ if the balls are disjoint, the center of each ball belongs to A, each ball has diameter at most ϵ. Note that if $\epsilon' < \epsilon''$, then an ϵ''-packing of A is also an ϵ'-packing. For $\alpha > 0$, the α-*dimensional packing pre-measure* on $\mathcal{X}$ assigns mass

$$P^\alpha(A) := \inf_{\epsilon > 0} \sup \left\{ \sum_i \operatorname{diam}(B_i)^\alpha \ : \ \{B_i\} \text{ is an } \epsilon\text{-packing of } A \right\}$$

to a set A. The α-*dimensional packing measure* on $\mathcal{X}$ is the Borel measure that assigns mass

$$\mathcal{P}^\alpha(A) := \inf\left\{\sum_i P^\alpha(A_i) \; : \; A \subseteq \bigcup_i A_i\right\}$$

to a Borel set A where the infimum is over all countable collections of Borel sets $\{A_i\}$ such that $A \subseteq \bigcup_i A_i$. The *packing dimension* of A is the infimum of those α such that the corresponding α-dimensional packing measure is zero.

Theorem B.1. *The packing dimension of a set is always at least as great as its Hausdorff dimension.*

We refer the reader to [107] for more about and properties of Hausdorff and packing dimension.

B.2 Energy and Capacity

Let $(\mathcal{X}, \rho)$ be a compact metric space. Write $M_1(\mathcal{X})$ for the collection of Borel probability measures on $\mathcal{X}$. A *gauge* is a function $f : [0, \infty[\to [0, \infty]$, such that:

- f is continuous and non-increasing,
- $f(0) = \infty$,
- $f(r) < \infty$ for $r > 0$,
- $\lim_{r \to \infty} f(r) = 0$.

Given $\mu \in M_1(\mathcal{X})$ and a gauge f, the *energy of μ in the gauge f* is the quantity

$$\mathcal{E}_f(\mu) := \int \mu(dx) \int \mu(dy) \, f(\rho(x,y)).$$

The *capacity of $\mathcal{X}$ in the gauge f* is the quantity

$$\mathrm{Cap}_f(\mathcal{X}) := (\inf\{\mathcal{E}_f(\mu) : \mu \in M_1(\mathcal{X})\})^{-1}$$

(note by our assumptions on f that we need only consider diffuse $\mu \in M_1(\mathcal{X})$ in the infimum).

The *capacity dimension* of $\mathcal{X}$ is the supremum of those $\alpha > 0$ such that $\mathcal{X}$ has strictly positive capacity in the gauge $f(x) = x^{-\alpha}$ (where we adopt the convention that the supremum of the empty set is 0).

Theorem B.2. *The Hausdorff and capacity dimensions of a compact metric space always coincide.*

We again refer to [107] for more about capacities and their connection to Hausdorff dimension.

B.3 Application to Trees from Coalescing Partitions

Recall the construction in Example 3.41 of a $\mathbb{R}$-tree and an associated ultra-metric completion $(\mathbb{S}, \delta)$ of $\mathbb{N}$ from a coalescing family $\{\Pi(t)\}_{t>0}$ of partitions of $\mathbb{N}$. We will assume that $\Pi(t)$ has finitely many blocks for $t > 0$, so that $(\mathbb{S}, \delta)$ is compact.

Write $N(t)$ for the number of blocks of $\Pi(t)$ and for $k \in \mathbb{N}$ put $\sigma_k := \inf\{t \geqslant 0 : N(t) \leqslant k\}$. The non-increasing function Π is constant on each of the intervals $[\sigma_k, \sigma_{k-1}[$, $k > 1$. Write $1 = I_1(t) < \cdots < I_{N(t)}(t)$ for an ordered listing of the least elements of the various blocks of $\Pi(t)$.

We can associate each partition $\Pi(t)$ with an equivalence relation $\sim_{\Pi(t)}$ on $\mathbb{N}$ by declaring that $i \sim_{\Pi(t)} j$ if i and j are in the same block of $\Pi(t)$.

Given $B \subseteq \mathbb{S}$, write $\mathrm{cl}B$ for the closure of B. Each of the sets

$$U_i(t) = \mathrm{cl}\{j \in \mathbb{N} : j \sim_{\Pi(t)} I_i(t)\}$$
$$= \mathrm{cl}\{j \in \mathbb{N} : \delta(j, I_i(t)) \leqslant 2t\}$$
$$= \{y \in \mathbb{S} : \delta(y, I_i(t)) \leqslant 2t\}$$

is a closed ball with diameter at most t (in an ultrametric space, the diameter and radius of a ball are equal). The closed balls of $\mathbb{S}$ are also the open balls and every ball is of the form $U_i(t)$ for some $t > 0$ – see, for example, Proposition 18.4 of [123] – and, in fact, every ball is of the form $U_i(\sigma_k)$ for some $k \in \mathbb{N}$ and $1 \leqslant i \leqslant k$. In particular, the collection of balls is countable. Any ball of diameter at most $2t$ is contained in a unique one of the $U_i(t)$, and any ball of diameter at least $2t$ contains one or more of the $U_i(t)$ – see, for example, Proposition 18.5 of [123].

We need to adapt to our setting the alternative expression for energy obtained by summation–by–parts in Section 2 of [112]. For $t > 0$ write $\mathcal{U}(t)$ for the collection of balls $\{U_1(t), \ldots, U_{N(t)}(t)\}$. Let $\mathcal{U}$ denote the union of these collections over all $t > 0$, so that $\mathcal{U}$ is just the countable collection of all balls of $\mathbb{S}$. Given $U \in \mathcal{U}$ with $U \neq \mathbb{S}$, let $U^{\rightarrow}$ denote the unique element of $\mathcal{U}$ such that there exists no $V \in \mathcal{U}$ with $U \subsetneq V \subsetneq U^{\rightarrow}$. More concretely, such a ball U is in $\mathcal{U}(\sigma_k)$ but not in $\mathcal{U}(\sigma_{k-1})$ for some unique $k > 1$, and $U^{\rightarrow}$ is the unique element of $\mathcal{U}(\sigma_{k-1})$ such that $U \subset U^{\rightarrow}$. Define $\mathbb{S}^{\rightarrow} := \dagger$, where $\dagger$ is an adjoined symbol. Put $\mathrm{diam}(\dagger) = \infty$.

Given a gauge f, write φ_f for the diffuse measure on $[0, \infty[$ such that $\varphi_f([r, \infty[) = \varphi_f(]r, \infty[) = f(r)$, $r \geqslant 0$. For a diffuse probability measure $\mu \in M_1(\mathbb{S})$ we have, with the convention $f(\infty) = 0$,

$$\mathcal{E}_f(\mu) = \int \mu(dx) \int \mu(dy)\, f(\delta(x,y))$$

$$= \int \mu(dx) \int \mu(dy) \sum_{U \in \mathcal{U},\, \{x,y\} \subseteq U} f(\mathrm{diam}(U)) - f(\mathrm{diam}(U^{\rightarrow}))$$

$$= \sum_{U \in \mathcal{U}} (f(\mathrm{diam}(U)) - f(\mathrm{diam}(U^{\rightarrow})))$$

$$\times \int \mu(dx) \int \mu(dy)\, \mathbf{1}\{\{x,y\} \subseteq U\} \tag{B.1}$$

$$= \sum_{U \in \mathcal{U}} (f(\mathrm{diam}(U)) - f(\mathrm{diam}(U^{\rightarrow})))\, \mu(U)^2$$

$$= \sum_{U \in \mathcal{U}} \int_{[0,\infty[} \varphi_f(dt)\, \mathbf{1}\{U \in \mathcal{U}(t)\} \mu(U)^2$$

$$= \int_{[0,\infty[} \varphi_f(dt) \sum_{U \in \mathcal{U}(t)} \mu(U)^2.$$

Proposition B.3. *Suppose for all* $t > 0$ *that the asymptotic block frequencies*

$$F_i(t) := \lim_{n\to\infty} n^{-1} \left| \{0 \leqslant j \leqslant n-1 : j \sim_{\Pi(t)} I_i(t)\} \right|,\ 1 \leqslant i \leqslant N(t),$$

exist and

$$F_1(t) + \cdots + F_{N(t)}(t) = 1.$$

Suppose also that for some $\alpha > 0$ *that*

$$0 < \liminf_{t\downarrow 0} t^\alpha N(t) \leqslant \limsup_{t\downarrow 0} t^\alpha N(t) < \infty$$

and

$$0 < \liminf_{t\downarrow 0} t^{-\alpha} \sum_{i=1}^{N(t)} F_i(t)^2 \leqslant \limsup_{t\downarrow 0} t^{-\alpha} \sum_{i=1}^{N(t)} F_i(t)^2 < \infty.$$

Then the Hausdorff and packing dimensions of $\mathbb{S}$ *are both* α *and there are constants* $0 < c' \leqslant c'' < \infty$ *such that for any gauge* f

$$c' \left(\int_0^1 f(t) t^{\alpha-1}\, dt \right)^{-1} \leqslant \mathrm{Cap}_f(\mathbb{S}) \leqslant c'' \left(\int_0^1 f(t) t^{\alpha-1}\, dt \right)^{-1}.$$

Proof. In order to establish that both the Hausdorff and packing dimensions of $\mathbb{S}$ are at most α it suffices to consider the packing dimension, because packing dimension always dominates Hausdorff dimension.

By definition of packing dimension, in order to establish that the packing dimension is at most α it suffices to show for each $\eta > \alpha$ that there is a constant $c < \infty$ such that for any packing $B_1, B_2, \ldots$ of $\mathbb{S}$ with balls of diameter at most 1, we have $\sum_k \mathrm{diam}(B_k)^\eta \leqslant c$. If $2 \cdot 2^{-p} \leqslant \mathrm{diam}(B_k) < 2 \cdot 2^{-(p-1)}$ for some $p \in \{0, 1, 2, \ldots\}$, then B_k contains one or more of the balls $U_i(2^{-p})$. Thus

$$|\{k \in \mathbb{N} : 2 \cdot 2^{-p} \leqslant \operatorname{diam}(B_k) < 2 \cdot 2^{-(p-1)}\}| \leqslant N(2^{-p})$$

and

$$\sum_k \operatorname{diam}(B_k)^\eta \leqslant \sum_{p=0}^{\infty} N(2^{-p})2^{-(p-1)\eta} < \infty,$$

as required.

If we establish the claim regarding capacities, then this will establish that the capacity dimension of $\mathbb{S}$ is α. This then gives the required lower bound on the packing and Hausdorff dimensions because the Hausdorff measure equals the capacity dimension and the packing dimension dominates Hausdorff dimension.

In order to establish the claimed lower bound on $\operatorname{Cap}_f(\mathbb{S})$ it appears, a priori, that for each gauge f we might need to find a probability measure μ depending on f such that $(\mathcal{E}_f(\mu))^{-1}$ is at least the left–hand side of the inequality. It turns out, however, that we can find a measure that works simultaneously for all gauges f. We construct this measure as follows.

Let $\mathcal{A}$ denote the algebra of subsets of $\mathbb{S}$ generated by the collection of balls $\mathcal{U}$. Thus, $\mathcal{A}$ is just the countable collection of finite unions of balls. The σ–algebra generated by $\mathcal{A}$ is the Borel σ–algebra of $\mathbb{S}$. The sets in $\mathcal{A}$ are compact, and, moreover, for all $k \in \mathbb{N}$ and indices $1 \leqslant i \leqslant k$ if $U_i(\sigma_k) = U_{i_1}(\sigma_{k+1}) \cup U_{i_2}(\sigma_{k+1}) \cup \cdots \cup U_{i_m}(\sigma_{k+1})$ (that is, if $\{I_{i_1}(\sigma_{k+1}), I_{i_2}(\sigma_{k+1}), \ldots, I_{i_m}(\sigma_{k+1})\} = \{I_\ell(\sigma_{k+1}) : I_\ell(\sigma_{k+1}) \sim_{\Pi(\sigma_k)} I_i(\sigma_k)\}$), then $F_i(\sigma_k) = F_{i_1}(\sigma_{k+1}) + F_{i_2}(\sigma_{k+1}) + \cdots + F_{i_m}(\sigma_{k+1})$. It is, therefore, possible to define a finitely additive set function ν on $\mathcal{A}$ such that

$$\nu(U_i(t)) = F_i(t), \; t > 0, \; 1 \leqslant i \leqslant N(t), \tag{B.2}$$

and

$$\nu(\mathbb{S}) = 1. \tag{B.3}$$

Furthermore, if $A_1 \supseteq A_2 \supseteq \ldots$ is a decreasing sequence of sets in the algebra $\mathcal{A}$ such that $\bigcap_n A_n = \emptyset$, then, by compactness, $A_n = \emptyset$ for all n sufficiently large and it is certainly the case that $\lim_{n\to\infty} \nu(A_n) = 0$. A standard extension theorem – see, for example, Theorems 3.1.1 and 3.1.4 of [48] – gives that the set function ν extends to a probability measure (also denoted by ν) on the Borel σ–algebra of $\mathbb{S}$.

From (B.1) we see that for some constant $0 < c' < \infty$ (not depending on f) we have

$$\mathrm{Cap}_f(\mathbb{S}) \geqslant (\mathcal{E}_f(\nu))^{-1} = \left(\int \varphi_f(dt) \sum_{U \in \mathcal{U}(t)} \nu(U)^2 \right)^{-1}$$

$$= \left(\int \varphi_f(dt) \sum_{i=1}^{N(t)} F_i(t)^2 \right)^{-1}$$

$$\geqslant c' \left(\int \varphi_f(dt)(t \wedge 1)^\alpha \right)^{-1} = c' \left(\int_0^1 f(t) t^{\alpha-1}\, dt \right)^{-1}.$$

Turning to the upper bound on $\mathrm{Cap}_f(\mathbb{S})$, note from the Cauchy-Schwarz inequality that for any $\mu \in M_1(\mathbb{S})$

$$1 = \left(\sum_{U \in \mathcal{U}(t)} \mu(U) \right)^2 \leqslant N(t) \sum_{U \in \mathcal{U}(t)} \mu(U)^2,$$

and so, by (B.1),

$$\mathrm{Cap}_f(\mathbb{S}) \leqslant \left(\int \varphi_f(dt) N(t)^{-1} \right)^{-1}$$

$$\leqslant c'' \left(\int \varphi_f(dt)(t \wedge 1)^\alpha \right)^{-1} = c'' \left(\int_0^1 f(t) t^{\alpha-1}\, dt \right)^{-1},$$

for some constant $0 < c'' < \infty$. □

Remark B.4. By the Cauchy-Schwarz inequality,

$$1 = \left(\sum_{i=1}^{N(t)} F_i(t) \right)^2 \leqslant \left(\sum_{i=1}^{N(t)} F_i(t)^2 \right) N(t).$$

Thus,

$$\limsup_{t \downarrow 0} t^\alpha N(t) < \infty \Longrightarrow \liminf_{t \downarrow 0} t^{-\alpha} \sum_{i=1}^{N(t)} F_i(t)^2 > 0$$

and

$$\limsup_{t \downarrow 0} t^{-\alpha} \sum_{i=1}^{N(t)} F_i(t)^2 < \infty \Longrightarrow \liminf_{t \downarrow 0} t^\alpha N(t) > 0.$$

References

1. Romain Abraham and Laurent Serlet. Poisson snake and fragmentation. *Electron. J. Probab.*, 7:no. 17, 15 pp. (electronic), 2002.
2. S. Albeverio and X. Zhao. On the relation between different constructions of random walks on p-adics. *Markov Process. Related Fields*, 6(2):239–255, 2000.
3. Sergio Albeverio. Theory of Dirichlet forms and applications. In *Lectures on probability theory and statistics (Saint-Flour, 2000)*, volume 1816 of *Lecture Notes in Math.*, pages 1–106. Springer, Berlin, 2003.
4. Sergio Albeverio and Witold Karwowski. Diffusion on p-adic numbers. In *Gaussian random fields (Nagoya, 1990)*, volume 1 of *Ser. Probab. Statist.*, pages 86–99. World Sci. Publ., River Edge, NJ, 1991.
5. Sergio Albeverio and Witold Karwowski. A random walk on p-adics—the generator and its spectrum. *Stochastic Process. Appl.*, 53(1):1–22, 1994.
6. Sergio Albeverio and Witold Karwowski. Real time random walks on p-adic numbers. In *Mathematical physics and stochastic analysis (Lisbon, 1998)*, pages 54–67. World Sci. Publ., River Edge, NJ, 2000.
7. Sergio Albeverio, Witold Karwowski, and Xuelei Zhao. Asymptotics and spectral results for random walks on p-adics. *Stochastic Process. Appl.*, 83(1):39–59, 1999.
8. Sergio Albeverio and Xuelei Zhao. A decomposition theorem for Lévy processes on local fields. *J. Theoret. Probab.*, 14(1):1–19, 2001.
9. Sergio Albeverio and Xuelei Zhao. A remark on nonsymmetric stochastic processes on p-adics. *Stochastic Anal. Appl.*, 20(2):243–261, 2002.
10. D. Aldous. The continuum random tree III. *Ann. Probab.*, 21:248–289, 1993.
11. D. J. Aldous. Exchangeability and related topics. In *École d'été de probabilités de Saint–Flour, XIII – 1983*, volume 1117 of *Lecture Notes in Math.*, pages 1–198. Springer, Berlin – New York, 1985.
12. David Aldous. The continuum random tree I. *Ann. Probab.*, 19:1–28, 1991.
13. David Aldous. The continuum random tree. II. An overview. In *Stochastic analysis (Durham, 1990)*, volume 167 of *London Math. Soc. Lecture Note Ser.*, pages 23–70. Cambridge Univ. Press, Cambridge, 1991.
14. David Aldous. Tree-valued Markov chains and Poisson-Galton-Watson distributions. In *Microsurveys in discrete probability (Princeton, NJ, 1997)*, volume 41 of *DIMACS Ser. Discrete Math. Theoret. Comput. Sci.*, pages 1–20. Amer. Math. Soc., Providence, RI, 1998.

15. David Aldous and Steven N. Evans. Dirichlet forms on totally disconnected spaces and bipartite Markov chains. *J. Theoret. Probab.*, 12(3):839–857, 1999.

16. David Aldous and Jim Pitman. Tree-valued Markov chains derived from Galton-Watson processes. *Ann. Inst. H. Poincaré Probab. Statist.*, 34(5):637–686, 1998.

17. David J. Aldous. The random walk construction of uniform spanning trees and uniform labelled trees. *SIAM J. Discrete Math.*, 3(4):450–465, 1990.

18. D.J. Aldous. Deterministic and stochastic models for coalescence (aggregation, coagulation): a review of the mean–field theory for probabilists. *Bernoulli*, 5:3–48, 1999.

19. Benjamin L. Allen and Mike Steel. Subtree transfer operations and their induced metrics on evolutionary trees. *Ann. Comb.*, 5(1):1–15, 2001.

20. J.M. Alonso, T. Brady, D. Cooper, V. Ferlini, M. Lustig, M. Mihalik, M. Shapiro, and H. Short. Notes on word hyperbolic groups. In H. Short, editor, *Group theory from a geometrical viewpoint (Trieste, 1990)*, pages 3–63. World Sci. Publishing, River Edge, NJ, 1991.

21. V. Anantharam and P. Tsoucas. A proof of the Markov chain tree theorem. *Statist. Probab. Lett.*, 8(2):189–192, 1989.

22. R. A. Arratia. *Coalescing Brownian motions on the line*. PhD thesis, University of Wisconsin–Madison, 1979.

23. S. Asmussen and T. G. Kurtz. Necessary and sufficient conditions for complete convergence in the law of large numbers. *Ann. Probab.*, 8(1):176–182, 1980.

24. Martin T. Barlow and Steven N. Evans. Markov processes on vermiculated spaces. In *Random walks and geometry*, pages 337–348. Walter de Gruyter GmbH & Co. KG, Berlin, 2004.

25. M.T. Barlow, M. Émery, F.B. Knight, S. Song, and M. Yor. Autour d'un théorème de Tsirelson sur les filtrations browniennes et non browniennes. In *Séminaire de Probabilités, XXXII*, volume 1686 of *Lecture Notes in Mathematics*, pages 264–305. Springer, Berlin, 1998.

26. M.T. Barlow, J. Pitman, and M. Yor. On Walsh's Brownian motions. In *Séminaire de Probabilités, XXXII*, volume 1372 of *Lecture Notes in Mathematics*, pages 275–293. Springer, Berlin, New York, 1989.

27. I. Benjamini and Y. Peres. Random walks on a tree and capacity in the interval. *Ann. Inst. H. Poincaré Probab. Statist.*, 28:557–592, 1992.

28. Julien Berestycki, Nathanael Berestycki, and Jason Schweinsberg. Beta-coalescents and continuous stable random trees.

29. Jean Bertoin. *Lévy processes*, volume 121 of *Cambridge Tracts in Mathematics*. Cambridge University Press, Cambridge, 1996.

30. Mladen Bestvina. ℝ-trees in topology, geometry, and group theory. In *Handbook of geometric topology*, pages 55–91. North-Holland, Amsterdam, 2002.

31. Philippe Biane, Jim Pitman, and Marc Yor. Probability laws related to the Jacobi theta and Riemann zeta functions, and Brownian excursions. *Bull. Amer. Math. Soc. (N.S.)*, 38(4):435–465 (electronic), 2001.

32. M. Bramson and D. Griffeath. Clustering and dispersion rates for some interacting particle systems on $\mathbb{Z}^1$. *Ann. Probab.*, 8:183–213, 1980.

33. Martin R. Bridson and André Haefliger. *Metric spaces of non-positive curvature*, volume 319 of *Grundlehren der Mathematischen Wissenschaften [Fundamental Principles of Mathematical Sciences]*. Springer-Verlag, Berlin, 1999.

34. Martin R. Bridson and Andre Haeflinger. *Metric Spaces and Non-Positive Curvature*. Springer, 2001.

35. A. Broder. Generating random spanning trees. In *Proc. 30'th IEEE Symp. Found. Comp. Sci.*, pages 442–447, 1989.

36. Peter Buneman. A note on the metric properties of trees. *J. Combinatorial Theory Ser. B*, 17:48–50, 1974.

37. Dmitri Burago, Yuri Burago, and Sergei Ivanov. *A course in metric geometry*, volume 33 of *Graduate studies in mathematics*. AMS, Boston, MA, 2001.

38. P. Cartier. Fonctions harmoniques sur un arbre. In *Convegno di Calcolo delle Probabilità, INDAM, Rome, 1971*, volume IX of *Symposia Mathematica*, pages 203–270. Academic Press, London, 1972.

39. Ian Chiswell. *Introduction to Λ-trees*. World Scientific Publishing Co. Inc., River Edge, NJ, 2001.

40. M. Coornaert, T. Delzant, and A. Papadopoulos. *Géométrie et théorie des groupes*, volume 1441 of *Lecture Notes in Mathematics*. Springer, 1990.

41. D. J. Daley and D. Vere-Jones. *An introduction to the theory of point processes*. Springer Series in Statistics. Springer-Verlag, New York, 1988.

42. D.S. Dean and K.M. Jansons. Brownian excursions on combs. *J. Statist. Phys.*, 70:1313–1332, 1993.

43. Claude Dellacherie and Paul-André Meyer. *Probabilities and potential. C*, volume 151 of *North-Holland Mathematics Studies*. North-Holland Publishing Co., Amsterdam, 1988. Potential theory for discrete and continuous semigroups, Translated from the French by J. Norris.

44. Peter Donnelly, Steven N. Evans, Klaus Fleischmann, Thomas G. Kurtz, and Xiaowen Zhou. Continuum-sites stepping-stone models, coalescing exchangeable partitions and random trees. *Ann. Probab.*, 28(3):1063–1110, 2000.

45. Andreas Dress, Vincent Moulton, and Werner Terhalle. *T*-theory: an overview. *European J. Combin.*, 17(2-3):161–175, 1996. Discrete metric spaces (Bielefeld, 1994).

46. Andreas W.M. Dress. Trees, tight extensions of metric spaces, and the cohomological dimension of certain groups: A note on combinatorical properties of metric spaces. *Adv. Math.*, 53:321–402, 1984.

47. A.W. Dress and W.F. Terhalle. The real tree. *Adv. Math.*, 120:283–301, 1996.

48. R.M. Dudley. *Real Analysis and Probability*. Wadsworth, Belmont CA, 1989.

49. Thomas Duquesne and Jean-François Le Gall. Probabilistic and fractal aspects of Lévy trees. *Probab. Theory Related Fields*, 131(4):553–603, 2005.

50. Thomas Duquesne and Matthias Winkel. Growth of Levy trees.

51. E.B. Dynkin. Integral representations of excessive measures and excessive functions. *Russian Math. Surveys*, 27:43–84, 1972.

52. E.B. Dynkin and M.B. Malyutov. Random walk on groups with a finite number of generators (Russian). *Dokl. Akad. Nauk SSSR*, 137:1042–1045, 1961.

53. N. Eisenbaum and H. Kaspi. A counterexample for the Markov property of local time for diffusions on graphs. In *Séminaire de Probabilités, XXIX*, volume 1613 of *Lecture Notes in Mathematics*, pages 260–265. Springer, Berlin, 1995.

54. N. Eisenbaum and H. Kaspi. On the Markov property of local time for Markov processes on graphs. *Stochastic Process. Appl.*, 64:153–172, 1996.

55. Alexei Ermakov, Bálint Tóth, and Wendelin Werner. On some annihilating and coalescing systems. *J. Statist. Phys.*, 91(5-6):845–870, 1998.

56. S.N. Ethier and T.G. Kurtz. *Markov Processes: Characterization and Convergence*. Wiley, New York, 1986.

57. Stewart N. Ethier and Thomas G. Kurtz. *Markov processes.* Wiley Series in Probability and Mathematical Statistics: Probability and Mathematical Statistics. John Wiley & Sons Inc., New York, 1986. Characterization and convergence.

58. S.N. Evans. Local field Gaussian measures. In E. Cinlar, K.L. Chung, and R.K. Getoor, editors, *Seminar on Stochastic Processes, 1988 (Gainesville, FL, 1988)*, pages 121–160. Birkhäuser, Boston, 1989.

59. S.N. Evans. Local properties of Lévy processes on a totally disconnected group. *J. Theoret. Probab.*, 2:209–259, 1989.

60. Steven N. Evans. Kingman's coalescent as a random metric space. In *Stochastic models (Ottawa, ON, 1998)*, volume 26 of *CMS Conf. Proc.*, pages 105–114. Amer. Math. Soc., Providence, RI, 2000.

61. Steven N. Evans. Snakes and spiders: Brownian motion on R-trees. *Probab. Theory Rel. Fields*, 117:361–386, 2000.

62. Steven N. Evans. Local fields, Gaussian measures, and Brownian motions. In *Topics in probability and Lie groups: boundary theory*, volume 28 of *CRM Proc. Lecture Notes*, pages 11–50. Amer. Math. Soc., Providence, RI, 2001.

63. Steven N. Evans, Jim Pitman, and Anita Winter. Rayleigh processes, real trees, and root growth with re-grafting. *Probab. Theory Related Fields*, 134(1):81–126, 2006.

64. Steven N. Evans and Richard B. Sowers. Pinching and twisting Markov processes. *Ann. Probab.*, 31(1):486–527, 2003.

65. Steven N. Evans and Anita Winter. Subtree prune and regraft: a reversible real tree-valued Markov process. *Ann. Probab.*, 34(3):918–961, 2006.

66. W. Feller. *An Introduction to Probability Theory and Its Applications*, volume II. Wiley, New York, 2nd edition, 1971.

67. Joseph Felsenstein. *Inferring Phylogenies.* Sinauer Associates, Sunderland, Massachusetts, 2003.

68. A. Figa-Talamanca. Diffusion on compact ultrametric spaces. In *Noncompact Lie Groups and some of their Applications (San Antonio, TX, 1993)*, pages 157–167. Kluwer, Dordrecht, 1994.

69. Peter Forster and Colin Renfrew, editors. *Phylogenetic methods and the prehistory of languages.* McDonald Institute for Archaeological Research, Cambridge, 2006.

70. D. Freedman. *Brownian Motion and Diffusion.* Springer, New York, 1983.

71. M.I. Freidlin and A.D. Wentzell. Diffusion processes on graphs and the averaging principle. *Ann. Probab.*, 21:2215–2245, 1993.

72. Masatoshi Fukushima, Yōichi Ōshima, and Masayoshi Takeda. *Dirichlet forms and symmetric Markov processes*, volume 19 of *de Gruyter Studies in Mathematics*. Walter de Gruyter & Co., Berlin, 1994.

73. M.A. Garcia Alvarez. Une théorie de la dualité à ensemble polaire près II. *Ann. Probab.*, 4:947–976, 1976.

74. M.A. Garcia Alvarez and P.A. Meyer. Une théorie de la dualité à ensemble polaire près I. *Ann. Probab.*, 1:207–222, 1973.

75. R.K. Getoor and J. Glover. Markov processes with identical excessive measures. *Math. Z.*, 184:287–300, 1983.

76. É. Ghys and P. de la Harpe, editors. *Sur les groupes hyperboliques d'apres Mikhael Gromov: papers from the Swiss seminar on hyperbolic groups held in Bern, 1988*, volume 83 of *Progress in Mathematics*. Birkhäuser, Boston, MA, 1990.

77. Stephen Jay Gould. *Dinosaur in a haystack: reflections in natural history.* Harmony Books, New York, 1995.

78. Andreas Greven, Peter Pfaffelhuber, and Anita Winter. Convergence in distribution of random metric measure spaces: (Λ-coalescent measure trees).

79. M. Gromov. Hyperbolic groups. In *Essays in group theory*, volume 8 of *Math. Sci. Res. Inst. Publ.*, pages 75–263. Springer, New York, 1987.

80. Misha Gromov. *Metric structures for Riemannian and non-Riemannian spaces*, volume 152 of *Progress in Mathematics*. Birkhäuser Boston Inc., Boston, MA, 1999. Based on the 1981 French original [MR 85e:53051], With appendices by M. Katz, P. Pansu and S. Semmes, Translated from the French by Sean Michael Bates.

81. Benedicte Haas, Gregory Miermont, Jim Pitman, and Matthias Winkel. Continuum tree asymptotics of discrete fragmentations and applications to phylogenetic models.

82. J. Hawkes. Trees generated by a simple branching process. *J. London Math. Soc. (2)*, 24:374–384, 1981.

83. Jotun Hein, Mikkel H. Schierup, and Carsten Wiuf. *Gene genealogies, variation and evolution.* Oxford University Press, Oxford, 2005. A primer in coalescent theory.

84. Juha Heinonen and Stephen Semmes. Thirty-three yes or no questions about mappings, measures, and metrics. *Conform. Geom. Dyn.*, 1:1–12 (electronic), 1997.

85. J.G. Hocking and G.S. Young. *Topology.* Addison–Wesley, Reading, MA, 1961.

86. T. Jeulin. Compactification de Martin d'un processus droit. *Z. Wahrscheinlichkeitstheorie verw. Gebiete*, 42:229–260, 1978.

87. Hiroshi Kaneko and Xuelei Zhao. Stochastic processes on $\mathbb{Q}_p$ induced by maps and recurrence criteria. *Forum Math.*, 16(1):69–95, 2004.

88. John L. Kelley. *General topology.* Springer-Verlag, New York, 1975. Reprint of the 1955 edition [Van Nostrand, Toronto, Ont.], Graduate Texts in Mathematics, No. 27.

89. J.F.C. Kingman. On the genealogy of large populations. In J. Gani and E.J. Hannan, editors, *Essays in Statistical Science*, pages 27–43. Applied Probability Trust, 1982. Special vol. 19A of *J. Appl. Probab.*

90. J.F.C. Kingman. The coalescent. *Stochastic Process. Appl.*, 13:235–248, 1982.

91. F.B. Knight. *Essentials of Brownian Motion and Diffusion*, volume 18 of *Mathematical Surveys and Monographs*. American Mathematical Society, Providence, 1981.

92. Frank B. Knight. *Essentials of Brownian motion and diffusion*, volume 18 of *Mathematical Surveys*. American Mathematical Society, Providence, R.I., 1981.

93. W.B. Krebs. Brownian motion on the continuum tree. *Probab. Theory Related Fields*, 101:421–433, 1995.

94. H. Kunita and T. Watanabe. Markov processes and Martin boundaries, I. *Illinois J. Math.*, 9:485–526, 1965.

95. T. J. Laakso. Ahlfors Q-regular spaces with arbitrary $Q > 1$ admitting weak Poincaré inequality. *Geom. Funct. Anal.*, 10(1):111–123, 2000.

96. Steven P. Lalley and Thomas Sellke. An extension of Hawkes' theorem on the Hausdorff dimension of a Galton-Watson tree. *Probab. Theory Related Fields*, 116(1):41–56, 2000.

97. J.-F. Le Gall. A class of path–valued Markov processes and its applications to superprocesses. *Probab. Theory Related Fields*, 95:25–46, 1993.

98. J.-F. Le Gall. A path-valued Markov process and its connections with partial differential equations. In *First European Congress of Mathematics, Vol. II (Paris, 1992)*, volume 120 of *Progr. Math.*, pages 185–212. Birkhuser, Basel, 1994.

99. J.-F. Le Gall. Hitting probabilities and potential theory for the Brownian path-valued process. *Ann. Inst. Fourier (Grenoble)*, 44:277–306, 1994.

100. J.-F. Le Gall. The Brownian snake and solutions of $\Delta u = u^2$ in a domain. *Probab. Theory Related Fields*, 102:393–432, 1995.

101. Jean-François Le Gall. Random trees and applications. *Probab. Surv.*, 2:245–311 (electronic), 2005.

102. Jean-François Le Gall. Random real trees. *Ann. Fac. Sci. Toulouse Math. (6)*, 15(1):35–62, 2006.

103. Jean-François Le Gall and Mathilde Weill. Conditioned Brownian trees. *Ann. Inst. H. Poincaré Probab. Statist.*, 42(4):455–489, 2006.

104. R.D. Lyons. Random walks and percolation on trees. *Ann. Probab.*, 18:931–958, 1990.

105. R.D. Lyons and Y. Peres. Probability on trees and networks. Book in preparation for Cambridge University Press, available via http://php.indiana.edu/~rdlyons/, 1996.

106. Z.-M. Ma and M. Röckner. *Introduction to the Theory of (Non–Symmetric) Dirichlet Forms*. Springer, Berlin, 1992.

107. P. Mattila. *Geometry of Sets and Measures in Euclidean Spaces: Fractals and Rectifiability*, volume 44 of *Cambridge Studies in Advanced Mathematics*. Cambridge University Press, Cambridge – New York, 1995.

108. P.-A. Meyer. *Processus de Markov: la Frontière de Martin*, volume 77 of *Lecture Notes in Mathematics*. Springer, Berlin, 1970.

109. M. Mezard, G. Parisi, and M.A. Virasoro. *Spin Glass Theory and Beyond*, volume 9 of *World Scientific Lecture Notes in Physics*. World Scientific, Singapore, 1987.

110. John W. Morgan. Λ-trees and their applications. *Bull. Amer. Math. Soc. (N.S.)*, 26(1):87–112, 1992.

111. Natella V. O'Bryant. A noisy system with a flattened Hamiltonian and multiple time scales. *Stoch. Dyn.*, 3(1):1–54, 2003.

112. R. Pemantle and Y. Peres. Galton–Watson trees with the same mean have the same polar sets. *Ann. Probab.*, 23:1102–1124, 1995.

113. R. Pemantle, Y. Peres, and J.W. Shapiro. The trace of spatial Brownian motion is capacity–equivalent to the unit square. *Probab. Theory Related Fields*, 106:379–399, 1996.

114. Y. Peres. Remarks on intersection–equivalence and capacity–equivalence. *Ann. Inst. H. Poincaré Phys. Théor.*, 64:339–347, 1996.

115. J. Pitman. *Combinatorial stochastic processes*, volume 1875 of *Lecture Notes in Mathematics*. Springer-Verlag, Berlin, 2006. Lectures from the 32nd Summer School on Probability Theory held in Saint-Flour, July 7–24, 2002, With a foreword by Jean Picard.

116. Jim Pitman. Coalescents with multiple collisions. *Ann. Probab.*, 27(4):1870–1902, 1999.

117. Daniel Revuz and Marc Yor. *Continuous martingales and Brownian motion*, volume 293 of *Grundlehren der Mathematischen Wissenschaften [Fundamental Principles of Mathematical Sciences]*. Springer-Verlag, Berlin, third edition, 1999.

118. D. Ringe, T. Warnow, and A. Taylor. Indo-European and computational cladistics. *Transactions of the Philological Society*, 100:59–129, 2002.

119. L. C. G. Rogers and David Williams. *Diffusions, Markov processes, and martingales. Vol. 2.* Cambridge Mathematical Library. Cambridge University Press, Cambridge, 2000. Itô calculus, Reprint of the second (1994) edition.

120. L.C.G. Rogers and D. Williams. *Diffusions, Markov Processes, and Martingales, Volume I: Foundations.* Wiley, 2nd edition, 1994.

121. H. L. Royden. *Real Analysis.* Collier MacMillan – New York, 2nd edition, 1968.

122. S. Sawyer. Isotropic random walks in a tree. *Z. Wahrsch. Verw. Gebiete*, 42:279–292, 1978.

123. W. H. Schikhof. *Ultrametric Calculus: an Introduction to p-adic Analysis*, volume 4 of *Cambridge Studies in Advanced Mathematics*. Cambridge University Press, Cambridge – New York, 1984.

124. D. Schwartz. On hitting probabilities for an annihilating particle model. *Ann. Probab.*, 6:398–403, 1976.

125. Charles Semple and Mike Steel. *Phylogenetics*, volume 24 of *Oxford Lecture Series in Mathematics and its Applications*. Oxford University Press, Oxford, 2003.

126. Peter B. Shalen. Dendrology of groups: an introduction. In *Essays in group theory*, volume 8 of *Math. Sci. Res. Inst. Publ.*, pages 265–319. Springer, New York, 1987.

127. Peter B. Shalen. Dendrology and its applications. In *Group theory from a geometrical viewpoint (Trieste, 1990)*, pages 543–616. World Sci. Publishing, River Edge, NJ, 1991.

128. M. Sharpe. *General Theory of Markov Processes.* Academic Press, San Diego, 1988.

129. G.F. Simmons. *Introduction to Topology and Modern Analysis.* McGraw–Hill, New York, 1963.

130. J. M. S. Simões Pereira. A note on the tree realizability of a distance matrix. *J. Combinatorial Theory*, 6:303–310, 1969.

131. Li Song, Li Jiao, and Xue Lei Zhao. Some time estimates of Lévy processes on p-adics. *J. Fudan Univ. Nat. Sci.*, 44(3):457–461, 476, 2005.

132. Florin Soucaliuc, Bálint Tóth, and Wendelin Werner. Reflection and coalescence between independent one-dimensional Brownian paths. *Ann. Inst. H. Poincaré Probab. Statist.*, 36(4):509–545, 2000.

133. Richard B. Sowers. Stochastic averaging with a flattened Hamiltonian: a Markov process on a stratified space (a whiskered sphere). *Trans. Amer. Math. Soc.*, 354(3):853–900 (electronic), 2002.

134. D.L. Swofford and G.J. Olsen. Phylogeny reconstruction. In D.M. Hillis and C. Moritz, editors, *Molecular Systematics*, pages 411–501. Sinauer Associates, Sunderland, Massachusetts, 1990.

135. M.H. Taibleson. *Fourier Analysis on Local Fields.* Princeton University Press, Princeton, N.J., 1975.

136. S. Tavaré. Line–of–descent and genealogical processes, and their applications in population genetics. *Theoret. Population Biol.*, 26:119–164, 1984.

137. W.F. Terhalle. ℝ-trees and symmetric differences of sets. *European J. Combin.*, 18:825–833, 1997.

138. Bálint Tóth and Wendelin Werner. The true self-repelling motion. *Probab. Theory Related Fields*, 111(3):375–452, 1998.

139. B. Tsirelson. Triple points: from non-brownian filtrations to harmonic measures. *Geom. Funct. Anal.*, 7:1096–1142, 1997.

140. B. Tsirelson. Brownian coalescence as a black noise I. Preprint, School of Mathematics, Tel Aviv University, 1998.

141. N.T. Varopoulos. Long range estimates for Markov chains. *Bull. Sc. Math.*, 109:225–252, 1985.

142. J.B. Walsh. A diffusion with discontinuous local time. In *Temps Locaux, Astérisque*, volume 52–53. Société Mathématique de France, Paris, 1978.

143. G.N. Watson. *A treatise on the theory of Bessel functions*. Cambridge University Press, Cambridge, second edition, 1944.

144. G. A. Watterson. Lines of descent and the coalescent. *Theoret. Population Biol.*, 26:77–92, 1984.

145. W. Woess. Random walks on infinite graphs and groups – a survey of selected topics. *Bull. London Math. Soc.*, 26:1–60, 1994.

146. Lorenzo Zambotti. A reflected stochastic heat equation as symmetric dynamics with respect to the 3-d Bessel bridge. *J. Funct. Anal.*, 180(1):195–209, 2001.

147. Lorenzo Zambotti. Integration by parts on Bessel bridges and related stochastic partial differential equations. *C. R. Math. Acad. Sci. Paris*, 334(3):209–212, 2002.

148. Lorenzo Zambotti. Integration by parts on δ-Bessel bridges, $\delta > 3$ and related SPDEs. *Ann. Probab.*, 31(1):323–348, 2003.

149. K. A. Zareckiĭ. Constructing a tree on the basis of a set of distances between the hanging vertices. *Uspehi Mat. Nauk*, 20(6):90–92, 1965.

Index

List of Participants

Lecturers

Ronald DONEY	Univ. Manchester, UK
Steven N. EVANS	Univ. California, Berkeley, USA
Cédric VILLANI	ENS Lyon, F

Participants

Larbi ALILI	Univ. Warwick, Coventry, UK
Sylvain ARLOT	Univ. Paris-Sud, Orsay, F
Fabrice BAUDOIN	Univ. Paul Sabatier, Toulouse, F
Hermine BIERMÉ	Univ. Orléans, F
François BOLLEY	ENS Lyon, F
Maria Emilia CABALERRO	Univ. Mexico
Francesco CARAVENNA	Univ. Pierre et Marie Curie, Paris, F
Loïc CHAUMONT	Univ. Pierre et Marie Curie, Paris, F
Charles CUTHBERTSON	Univ. Oxford, UK
Latifa DEBBI	Univ. Henri Poincaré, Nancy, F
Pierre DEBS	Univ. Henri Poincaré, Nancy, F
Jérôme DEMANGE	Univ. Paul Sabatier, Toulouse, F
Hacène DJELLOUT	Univ. Blaise Pascal, Clermont-Ferrand, F
Coralie DUBOIS	Univ. Claude Bernard, Lyon, F
Anne EYRAUD-LOISEL	Univ. Claude Bernard, Lyon, F
Neil FARRICKER	Univ. Manchester, UK

Uwe FRANZ	Inst. Biomath. Biometry, Neuherberg, D
Christina GOLDSCHMIDT	Univ. Cambridge, UK
Jean-Baptiste GOUÉRÉ	Univ. Claude Bernard, Lyon, F
Mathieu GOURCY	Univ. Blaise Pascal, Clermont-Ferrand, F
Priscilla GREENWOOD	Arizona State Univ., Tempe, USA
Bénédicte HAAS	Univ. Oxford, UK
Christopher HOWITT	Univ. Oxford, UK
Jérémie JAKUBOWICZ	ENS Cachan, F
Aldéric JOULIN	Univ. La Rochelle, F
Pawel KISOWSKI	Univ. Wroclaw, Poland
Nathalie KRELL	Univ. Pierre et Marie Curie, Paris, F
Aline KURTZMANN	Univ. Neuchâtel, Switzerland
Krzysztof LATUSZYŃSKI	Warsaw School Economics, Poland
Liangzhen LEI	Univ. Blaise Pascal, Clermont-Ferrand, F
Christophe LEURIDAN	Univ. J. Fourier, Grenoble, F
Stéphane LOISEL	Univ. Claude Bernard, Lyon, F
Jose Alfredo LOPEZ MIMBELA	CIMAT, Guanajuato, Mexico
Mike LUDKOVSKI	Princeton Univ., USA
Yutao MA	Univ. La Rochelle, F
Philippe MARCHAL	ENS Paris, F
James MARTIN	Univ. Paris 7, F
Marie-Amélie MORLAIS	Univ. Rennes 1, F
Jan OBLÓJ	Univ. Pierre et Marie Curie, Paris, F
Cyril ODASSO	Univ. Rennes 1, F
Juan Carlos PARDO MILLAN	Univ. Pierre et Marie Curie, Paris, F
Robert PHILIPOWSKI	Univ. Bonn, D
Jean PICARD	Univ. Blaise Pascal, Clermont-Ferrand, F
Victor RIVERO MERCADO	Univ. Paris 10, F
Erwan SAINT LOUBERT BIÉ	Univ. Blaise Pascal, Clermont-Ferrand, F
Catherine SAVONA	Univ. Blaise Pascal, Clermont-Ferrand, F
François SIMENHAUS	Univ. Pierre et Marie Curie, Paris, F
Tommi SOTTINEN	Univ. Helsinki, Finland

I. TORRECILLA-TARANTINO	Univ. Barcelona, Spain
Gerónimo URIBE	Univ. Mexico
Vincent VIGON	Univ. Strasbourg, F
Matthias WINKEL	Univ. Oxford, UK
Marcus WUNSCH	Univ. Wien, Austria

List of Short Lectures

Larbi Alili	On some functional transformations and an application to the boundary crossing problem for a Brownian motion
Fabrice Baudoin	Stochastic differential equations and differential operators
Hermine Biermé	Random fields: self-similarity, anisotropy and directional analysis
François Bolley	Approximation of some diffusion PDE by some interacting particle system
Francesco Caravenna	A renewal theory approach to periodically inhomogeneous polymer models
Loïc Chaumont	On positive self-similar Markov processes
Charles Cuthbertson	Multiple selective sweeps and multi-type branching
Jérôme Demange	Porous media equation and Sobolev inequalities
Anne Eyraud-Loisel	Backward and forward-backward stochastic differential equations with enlarged filtration
Neil Farricker	Spectrally negative Lévy processes
Uwe Franz	A probabilistic model for biological clocks

Robert Philipowski	Propagation du chaos pour l'équation des milieux poreux
Tommi Sottinen	On the equivalence of multiparameter Gaussian processes
Gerónimo Uribe	Markov bridges, backward times, and a Brownian fragmentation
Vincent Vigon	Certains comportements des processus de Lévy sont décryptables par la factorisation de Wiener-Hopf
Matthias Winkel	Coupling construction of Lévy trees
Marcus Wunsch	A stability result for drift-diffusion-Poisson systems

Saint-Flour Probability Summer Schools

In order to facilitate research concerning previous schools we give here the names of the authors, the series*, and the number of the volume where their lectures can be found:

Summer School	Authors	Series	Vol. Nr.
1971	Bretagnolle; Chatterji; Meyer	LNM	307
1973	Meyer; Priouret; Spitzer	LNM	390
1974	Fernique; Conze; Gani	LNM	480
1975	Badrikian; Kingman; Kuelbs	LNM	539
1976	Hoffmann-Jörgensen; Liggett; Neveu	LNM	598
1977	Dacunha-Castelle; Heyer; Roynette	LNM	678
1978	Azencott; Guivarc'h; Gundy	LNM	774
1979	Bickel; El Karoui; Yor	LNM	876
1980	Bismut; Gross; Krickeberg	LNM	929
1981	Fernique; Millar; Stroock; Weber	LNM	976
1982	Dudley; Kunita; Ledrappier	LNM	1097
1983	Aldous; Ibragimov; Jacod	LNM	1117
1984	Carmona; Kesten; Walsh	LNM	1180
1985/86/87	Diaconis; Elworthy; Föllmer; Nelson; Papanicolaou; Varadhan	LNM	1362
1986	Barndorff-Nielsen	LNS	50
1988	Ancona; Geman; Ikeda	LNM	1427
1989	Burkholder; Pardoux; Sznitman	LNM	1464
1990	Freidlin; Le Gall	LNM	1527
1991	Dawson; Maisonneuve; Spencer	LNM	1541
1992	Bakry; Gill; Molchanov	LNM	1581
1993	Biane; Durrett;	LNM	1608
1994	Dobrushin; Groeneboom; Ledoux	LNM	1648
1995	Barlow; Nualart	LNM	1690
1996	Giné; Grimmett; Saloff-Coste	LNM	1665
1997	Bertoin; Martinelli; Peres	LNM	1717
1998	Emery; Nemirovski; Voiculescu	LNM	1738
1999	Bolthausen; Perkins; van der Vaart	LNM	1781
2000	Albeverio; Schachermayer; Talagrand	LNM	1816
2001	Tavaré; Zeitouni	LNM	1837
	Catoni	LNM	1851
2002	Tsirelson; Werner	LNM	1840
	Pitman	LNM	1875
2003	Dembo; Funaki	LNM	1869
	Massart	LNM	1896
2004	Cerf	LNM	1878
	Slade	LNM	1879
	Lyons**, Caruana, Lévy	LNM	1908
2005	Doney	LNM	1897
	Evans	LNM	1920
	Villani	Forthcoming	

*Lecture Notes in Mathematics (LNM), Lecture Notes in Statistics (LNS)
**The St. Flour lecturer was T.J. Lyons.

Lecture Notes in Mathematics

For information about earlier volumes
please contact your bookseller or Springer
LNM Online archive: springerlink.com

Vol. 1781: E. Bolthausen, E. Perkins, A. van der Vaart, Lectures on Probability Theory and Statistics. Ecole d' Eté de Probabilités de Saint-Flour XXIX-1999. Editor: P. Bernard (2002)

Vol. 1782: C.-H. Chu, A. T.-M. Lau, Harmonic Functions on Groups and Fourier Algebras (2002)

Vol. 1783: L. Grüne, Asymptotic Behavior of Dynamical and Control Systems under Perturbation and Discretization (2002)

Vol. 1784: L. H. Eliasson, S. B. Kuksin, S. Marmi, J.-C. Yoccoz, Dynamical Systems and Small Divisors. Cetraro, Italy 1998. Editors: S. Marmi, J.-C. Yoccoz (2002)

Vol. 1785: J. Arias de Reyna, Pointwise Convergence of Fourier Series (2002)

Vol. 1786: S. D. Cutkosky, Monomialization of Morphisms from 3-Folds to Surfaces (2002)

Vol. 1787: S. Caenepeel, G. Militaru, S. Zhu, Frobenius and Separable Functors for Generalized Module Categories and Nonlinear Equations (2002)

Vol. 1788: A. Vasil'ev, Moduli of Families of Curves for Conformal and Quasiconformal Mappings (2002)

Vol. 1789: Y. Sommerhäuser, Yetter-Drinfel'd Hopf algebras over groups of prime order (2002)

Vol. 1790: X. Zhan, Matrix Inequalities (2002)

Vol. 1791: M. Knebusch, D. Zhang, Manis Valuations and Prüfer Extensions I: A new Chapter in Commutative Algebra (2002)

Vol. 1792: D. D. Ang, R. Gorenflo, V. K. Le, D. D. Trong, Moment Theory and Some Inverse Problems in Potential Theory and Heat Conduction (2002)

Vol. 1793: J. Cortés Monforte, Geometric, Control and Numerical Aspects of Nonholonomic Systems (2002)

Vol. 1794: N. Pytheas Fogg, Substitution in Dynamics, Arithmetics and Combinatorics. Editors: V. Berthé, S. Ferenczi, C. Mauduit, A. Siegel (2002)

Vol. 1795: H. Li, Filtered-Graded Transfer in Using Noncommutative Gröbner Bases (2002)

Vol. 1796: J.M. Melenk, hp-Finite Element Methods for Singular Perturbations (2002)

Vol. 1797: B. Schmidt, Characters and Cyclotomic Fields in Finite Geometry (2002)

Vol. 1798: W.M. Oliva, Geometric Mechanics (2002)

Vol. 1799: H. Pajot, Analytic Capacity, Rectifiability, Menger Curvature and the Cauchy Integral (2002)

Vol. 1800: O. Gabber, L. Ramero, Almost Ring Theory (2003)

Vol. 1801: J. Azéma, M. Émery, M. Ledoux, M. Yor (Eds.), Séminaire de Probabilités XXXVI (2003)

Vol. 1802: V. Capasso, E. Merzbach, B. G. Ivanoff, M. Dozzi, R. Dalang, T. Mountford, Topics in Spatial Stochastic Processes. Martina Franca, Italy 2001. Editor: E. Merzbach (2003)

Vol. 1803: G. Dolzmann, Variational Methods for Crystalline Microstructure – Analysis and Computation (2003)

Vol. 1804: I. Cherednik, Ya. Markov, R. Howe, G. Lusztig, Iwahori-Hecke Algebras and their Representation Theory. Martina Franca, Italy 1999. Editors: V. Baldoni, D. Barbasch (2003)

Vol. 1805: F. Cao, Geometric Curve Evolution and Image Processing (2003)

Vol. 1806: H. Broer, I. Hoveijn. G. Lunther, G. Vegter, Bifurcations in Hamiltonian Systems. Computing Singularities by Gröbner Bases (2003)

Vol. 1807: V. D. Milman, G. Schechtman (Eds.), Geometric Aspects of Functional Analysis. Israel Seminar 2000-2002 (2003)

Vol. 1808: W. Schindler, Measures with Symmetry Properties (2003)

Vol. 1809: O. Steinbach, Stability Estimates for Hybrid Coupled Domain Decomposition Methods (2003)

Vol. 1810: J. Wengenroth, Derived Functors in Functional Analysis (2003)

Vol. 1811: J. Stevens, Deformations of Singularities (2003)

Vol. 1812: L. Ambrosio, K. Deckelnick, G. Dziuk, M. Mimura, V. A. Solonnikov, H. M. Soner, Mathematical Aspects of Evolving Interfaces. Madeira, Funchal, Portugal 2000. Editors: P. Colli, J. F. Rodrigues (2003)

Vol. 1813: L. Ambrosio, L. A. Caffarelli, Y. Brenier, G. Buttazzo, C. Villani, Optimal Transportation and its Applications. Martina Franca, Italy 2001. Editors: L. A. Caffarelli, S. Salsa (2003)

Vol. 1814: P. Bank, F. Baudoin, H. Föllmer, L.C.G. Rogers, M. Soner, N. Touzi, Paris-Princeton Lectures on Mathematical Finance 2002 (2003)

Vol. 1815: A. M. Vershik (Ed.), Asymptotic Combinatorics with Applications to Mathematical Physics. St. Petersburg, Russia 2001 (2003)

Vol. 1816: S. Albeverio, W. Schachermayer, M. Talagrand, Lectures on Probability Theory and Statistics. Ecole d'Eté de Probabilités de Saint-Flour XXX-2000. Editor: P. Bernard (2003)

Vol. 1817: E. Koelink, W. Van Assche (Eds.), Orthogonal Polynomials and Special Functions. Leuven 2002 (2003)

Vol. 1818: M. Bildhauer, Convex Variational Problems with Linear, nearly Linear and/or Anisotropic Growth Conditions (2003)

Vol. 1819: D. Masser, Yu. V. Nesterenko, H. P. Schlickewei, W. M. Schmidt, M. Waldschmidt, Diophantine Approximation. Cetraro, Italy 2000. Editors: F. Amoroso, U. Zannier (2003)

Vol. 1820: F. Hiai, H. Kosaki, Means of Hilbert Space Operators (2003)

Vol. 1821: S. Teufel, Adiabatic Perturbation Theory in Quantum Dynamics (2003)

Vol. 1822: S.-N. Chow, R. Conti, R. Johnson, J. Mallet-Paret, R. Nussbaum, Dynamical Systems. Cetraro, Italy 2000. Editors: J. W. Macki, P. Zecca (2003)

Vol. 1823: A. M. Anile, W. Allegretto, C. Ringhofer, Mathematical Problems in Semiconductor Physics. Cetraro, Italy 1998. Editor: A. M. Anile (2003)

Vol. 1824: J. A. Navarro González, J. B. Sancho de Salas, $\mathscr{C}^\infty$ – Differentiable Spaces (2003)

Vol. 1825: J. H. Bramble, A. Cohen, W. Dahmen, Multiscale Problems and Methods in Numerical Simulations, Martina Franca, Italy 2001. Editor: C. Canuto (2003)

Vol. 1826: K. Dohmen, Improved Bonferroni Inequalities via Abstract Tubes. Inequalities and Identities of Inclusion-Exclusion Type. VIII, 113 p, 2003.

Vol. 1827: K. M. Pilgrim, Combinations of Complex Dynamical Systems. IX, 118 p, 2003.

Vol. 1828: D. J. Green, Gröbner Bases and the Computation of Group Cohomology. XII, 138 p, 2003.

Vol. 1829: E. Altman, B. Gaujal, A. Hordijk, Discrete-Event Control of Stochastic Networks: Multimodularity and Regularity. XIV, 313 p, 2003.

Vol. 1830: M. I. Gil', Operator Functions and Localization of Spectra. XIV, 256 p, 2003.

Vol. 1831: A. Connes, J. Cuntz, E. Guentner, N. Higson, J. E. Kaminker, Noncommutative Geometry, Martina Franca, Italy 2002. Editors: S. Doplicher, L. Longo (2004)

Vol. 1832: J. Azéma, M. Émery, M. Ledoux, M. Yor (Eds.), Séminaire de Probabilités XXXVII (2003)

Vol. 1833: D.-Q. Jiang, M. Qian, M.-P. Qian, Mathematical Theory of Nonequilibrium Steady States. On the Frontier of Probability and Dynamical Systems. IX, 280 p, 2004.

Vol. 1834: Yo. Yomdin, G. Comte, Tame Geometry with Application in Smooth Analysis. VIII, 186 p, 2004.

Vol. 1835: O.T. Izhboldin, B. Kahn, N.A. Karpenko, A. Vishik, Geometric Methods in the Algebraic Theory of Quadratic Forms. Summer School, Lens, 2000. Editor: J.-P. Tignol (2004)

Vol. 1836: C. Năstăsescu, F. Van Oystaeyen, Methods of Graded Rings. XIII, 304 p, 2004.

Vol. 1837: S. Tavaré, O. Zeitouni, Lectures on Probability Theory and Statistics. Ecole d'Eté de Probabilités de Saint-Flour XXXI-2001. Editor: J. Picard (2004)

Vol. 1838: A.J. Ganesh, N.W. O'Connell, D.J. Wischik, Big Queues. XII, 254 p, 2004.

Vol. 1839: R. Gohm, Noncommutative Stationary Processes. VIII, 170 p, 2004.

Vol. 1840: B. Tsirelson, W. Werner, Lectures on Probability Theory and Statistics. Ecole d'Eté de Probabilités de Saint-Flour XXXII-2002. Editor: J. Picard (2004)

Vol. 1841: W. Reichel, Uniqueness Theorems for Variational Problems by the Method of Transformation Groups (2004)

Vol. 1842: T. Johnsen, A. L. Knutsen, K₃ Projective Models in Scrolls (2004)

Vol. 1843: B. Jefferies, Spectral Properties of Noncommuting Operators (2004)

Vol. 1844: K.F. Siburg, The Principle of Least Action in Geometry and Dynamics (2004)

Vol. 1845: Min Ho Lee, Mixed Automorphic Forms, Torus Bundles, and Jacobi Forms (2004)

Vol. 1846: H. Ammari, H. Kang, Reconstruction of Small Inhomogeneities from Boundary Measurements (2004)

Vol. 1847: T.R. Bielecki, T. Björk, M. Jeanblanc, M. Rutkowski, J.A. Scheinkman, W. Xiong, Paris-Princeton Lectures on Mathematical Finance 2003 (2004)

Vol. 1848: M. Abate, J. E. Fornaess, X. Huang, J. P. Rosay, A. Tumanov, Real Methods in Complex and CR Geometry, Martina Franca, Italy 2002. Editors: D. Zaitsev, G. Zampieri (2004)

Vol. 1849: Martin L. Brown, Heegner Modules and Elliptic Curves (2004)

Vol. 1850: V. D. Milman, G. Schechtman (Eds.), Geometric Aspects of Functional Analysis. Israel Seminar 2002-2003 (2004)

Vol. 1851: O. Catoni, Statistical Learning Theory and Stochastic Optimization (2004)

Vol. 1852: A.S. Kechris, B.D. Miller, Topics in Orbit Equivalence (2004)

Vol. 1853: Ch. Favre, M. Jonsson, The Valuative Tree (2004)

Vol. 1854: O. Saeki, Topology of Singular Fibers of Differential Maps (2004)

Vol. 1855: G. Da Prato, P.C. Kunstmann, I. Lasiecka, A. Lunardi, R. Schnaubelt, L. Weis, Functional Analytic Methods for Evolution Equations. Editors: M. Iannelli, R. Nagel, S. Piazzera (2004)

Vol. 1856: K. Back, T.R. Bielecki, C. Hipp, S. Peng, W. Schachermayer, Stochastic Methods in Finance, Bressanone/Brixen, Italy, 2003. Editors: M. Fritelli, W. Runggaldier (2004)

Vol. 1857: M. Émery, M. Ledoux, M. Yor (Eds.), Séminaire de Probabilités XXXVIII (2005)

Vol. 1858: A.S. Cherny, H.-J. Engelbert, Singular Stochastic Differential Equations (2005)

Vol. 1859: E. Letellier, Fourier Transforms of Invariant Functions on Finite Reductive Lie Algebras (2005)

Vol. 1860: A. Borisyuk, G.B. Ermentrout, A. Friedman, D. Terman, Tutorials in Mathematical Biosciences I. Mathematical Neurosciences (2005)

Vol. 1861: G. Benettin, J. Henrard, S. Kuksin, Hamiltonian Dynamics – Theory and Applications, Cetraro, Italy, 1999. Editor: A. Giorgilli (2005)

Vol. 1862: B. Helffer, F. Nier, Hypoelliptic Estimates and Spectral Theory for Fokker-Planck Operators and Witten Laplacians (2005)

Vol. 1863: H. Führ, Abstract Harmonic Analysis of Continuous Wavelet Transforms (2005)

Vol. 1864: K. Efstathiou, Metamorphoses of Hamiltonian Systems with Symmetries (2005)

Vol. 1865: D. Applebaum, B.V. R. Bhat, J. Kustermans, J. M. Lindsay, Quantum Independent Increment Processes I. From Classical Probability to Quantum Stochastic Calculus. Editors: M. Schürmann, U. Franz (2005)

Vol. 1866: O.E. Barndorff-Nielsen, U. Franz, R. Gohm, B. Kümmerer, S. Thorbjønsen, Quantum Independent Increment Processes II. Structure of Quantum Lévy Processes, Classical Probability, and Physics. Editors: M. Schürmann, U. Franz, (2005)

Vol. 1867: J. Sneyd (Ed.), Tutorials in Mathematical Biosciences II. Mathematical Modeling of Calcium Dynamics and Signal Transduction. (2005)

Vol. 1868: J. Jorgenson, S. Lang, Posₙ(R) and Eisenstein Series. (2005)

Vol. 1869: A. Dembo, T. Funaki, Lectures on Probability Theory and Statistics. Ecole d'Eté de Probabilités de Saint-Flour XXXIII-2003. Editor: J. Picard (2005)

Vol. 1870: V.I. Gurariy, W. Lusky, Geometry of Müntz Spaces and Related Questions. (2005)

Vol. 1871: P. Constantin, G. Gallavotti, A.V. Kazhikhov, Y. Meyer, S. Ukai, Mathematical Foundation of Turbulent Viscous Flows, Martina Franca, Italy, 2003. Editors: M. Cannone, T. Miyakawa (2006)

Vol. 1872: A. Friedman (Ed.), Tutorials in Mathematical Biosciences III. Cell Cycle, Proliferation, and Cancer (2006)

Vol. 1873: R. Mansuy, M. Yor, Random Times and Enlargements of Filtrations in a Brownian Setting (2006)

Vol. 1874: M. Yor, M. Émery (Eds.), In Memoriam Paul-André Meyer - Séminaire de Probabilités XXXIX (2006)

Vol. 1875: J. Pitman, Combinatorial Stochastic Processes. Ecole d'Eté de Probabilités de Saint-Flour XXXII-2002. Editor: J. Picard (2006)

Vol. 1876: H. Herrlich, Axiom of Choice (2006)

Vol. 1877: J. Steuding, Value Distributions of L-Functions (2007)

Vol. 1878: R. Cerf, The Wulff Crystal in Ising and Percolation Models, Ecole d'Eté de Probabilités de Saint-Flour XXXIV-2004. Editor: Jean Picard (2006)

Vol. 1879: G. Slade, The Lace Expansion and its Applications, Ecole d'Eté de Probabilités de Saint-Flour XXXIV-2004. Editor: Jean Picard (2006)

Vol. 1880: S. Attal, A. Joye, C.-A. Pillet, Open Quantum Systems I, The Hamiltonian Approach (2006)

Vol. 1881: S. Attal, A. Joye, C.-A. Pillet, Open Quantum Systems II, The Markovian Approach (2006)

Vol. 1882: S. Attal, A. Joye, C.-A. Pillet, Open Quantum Systems III, Recent Developments (2006)

Vol. 1883: W. Van Assche, F. Marcellàn (Eds.), Orthogonal Polynomials and Special Functions, Computation and Application (2006)

Vol. 1884: N. Hayashi, E.I. Kaikina, P.I. Naumkin, I.A. Shishmarev, Asymptotics for Dissipative Nonlinear Equations (2006)
Vol. 1885: A. Telcs, The Art of Random Walks (2006)
Vol. 1886: S. Takamura, Splitting Deformations of Degenerations of Complex Curves (2006)
Vol. 1887: K. Habermann, L. Habermann, Introduction to Symplectic Dirac Operators (2006)
Vol. 1888: J. van der Hoeven, Transseries and Real Differential Algebra (2006)
Vol. 1889: G. Osipenko, Dynamical Systems, Graphs, and Algorithms (2006)
Vol. 1890: M. Bunge, J. Funk, Singular Coverings of Toposes (2006)
Vol. 1891: J.B. Friedlander, D.R. Heath-Brown, H. Iwaniec, J. Kaczorowski, Analytic Number Theory, Cetraro, Italy, 2002. Editors: A. Perelli, C. Viola (2006)
Vol. 1892: A. Baddeley, I. Bárány, R. Schneider, W. Weil, Stochastic Geometry, Martina Franca, Italy, 2004. Editor: W. Weil (2007)
Vol. 1893: H. Hanßmann, Local and Semi-Local Bifurcations in Hamiltonian Dynamical Systems, Results and Examples (2007)
Vol. 1894: C.W. Groetsch, Stable Approximate Evaluation of Unbounded Operators (2007)
Vol. 1895: L. Molnár, Selected Preserver Problems on Algebraic Structures of Linear Operators and on Function Spaces (2007)
Vol. 1896: P. Massart, Concentration Inequalities and Model Selection, Ecole d'Été de Probabilités de Saint-Flour XXXIII-2003. Editor: J. Picard (2007)
Vol. 1897: R. Doney, Fluctuation Theory for Lévy Processes, Ecole d'Été de Probabilités de Saint-Flour XXXV-2005. Editor: J. Picard (2007)
Vol. 1898: H.R. Beyer, Beyond Partial Differential Equations, On linear and Quasi-Linear Abstract Hyperbolic Evolution Equations (2007)
Vol. 1899: Séminaire de Probabilités XL. Editors: C. Donati-Martin, M. Émery, A. Rouault, C. Stricker (2007)
Vol. 1900: E. Bolthausen, A. Bovier (Eds.), Spin Glasses (2007)
Vol. 1901: O. Wittenberg, Intersections de deux quadriques et pinceaux de courbes de genre 1, Intersections of Two Quadrics and Pencils of Curves of Genus 1 (2007)
Vol. 1902: A. Isaev, Lectures on the Automorphism Groups of Kobayashi-Hyperbolic Manifolds (2007)
Vol. 1903: G. Kresin, V. Maz'ya, Sharp Real-Part Theorems (2007)
Vol. 1904: P. Giesl, Construction of Global Lyapunov Functions Using Radial Basis Functions (2007)
Vol. 1905: C. Prévôt, M. Röckner, A Concise Course on Stochastic Partial Differential Equations (2007)
Vol. 1906: T. Schuster, The Method of Approximate Inverse: Theory and Applications (2007)
Vol. 1907: M. Rasmussen, Attractivity and Bifurcation for Nonautonomous Dynamical Systems (2007)
Vol. 1908: T.J. Lyons, M. Caruana, T. Lévy, Differential Equations Driven by Rough Paths, Ecole d'Été de Probabilités de Saint-Flour XXXIV-2004 (2007)
Vol. 1909: H. Akiyoshi, M. Sakuma, M. Wada, Y. Yamashita, Punctured Torus Groups and 2-Bridge Knot Groups (I) (2007)
Vol. 1910: V.D. Milman, G. Schechtman (Eds.), Geometric Aspects of Functional Analysis. Israel Seminar 2004-2005 (2007)

Vol. 1911: A. Bressan, D. Serre, M. Williams, K. Zumbrun, Hyperbolic Systems of Balance Laws. Lectures given at the C.I.M.E. Summer School held in Cetraro, Italy, July 14–21, 2003. Editor: P. Marcati (2007)
Vol. 1912: V. Berinde, Iterative Approximation of Fixed Points (2007)
Vol. 1913: J.E. Marsden, G. Misiołek, J.-P. Ortega, M. Perlmutter, T.S. Ratiu, Hamiltonian Reduction by Stages (2007)
Vol. 1914: G. Kutyniok, Affine Density in Wavelet Analysis (2007)
Vol. 1915: T. Bıyıkoğlu, J. Leydold, P.F. Stadler, Laplacian Eigenvectors of Graphs. Perron-Frobenius and Faber-Krahn Type Theorems (2007)
Vol. 1916: C. Villani, F. Rezakhanlou, Entropy Methods for the Boltzmann Equation. Editors: F. Golse, S. Olla (forthcoming)
Vol. 1917: I. Veselić, Existence and Regularity Properties of the Integrated Density of States of Random Schrödinger (2007)
Vol. 1918: B. Roberts, R. Schmidt, Local Newforms for GSp(4) (2007)
Vol. 1919: R.A. Carmona, I. Ekeland, A. Kohatsu-Higa, J.-M. Lasry, P.-L. Lions, H. Pham, E. Taflin, Paris-Princeton Lectures on Mathematical Finance 2004. Editors: R.A. Carmona, E. Çinlar, I. Ekeland, E. Jouini, J.A. Scheinkman, N. Touzi (2007)
Vol. 1920: S.N. Evans, Probability and Real Trees. École d'Été de Probabilités de Saint-Flour XXXV-2005 (2008)
Vol. 1921: J.P. Tian, Evolution Algebras and their Applications (2008)
Vol. 1922: A. Friedman (Ed.), Tutorials in Mathematical BioSciences IV. Evolution and Ecology (2008)
Vol. 1923: J.P.N. Bishwal, Parameter Estimation in Stochastic Differential Equations (2008)
Vol. 1924: M. Wilson, Littlewood-Paley Theory and Exponential-Square Integrability (2008)
Vol. 1925: M. du Sautoy, Zeta Functions of Groups and Rings (2008)
Vol. 1926: L. Barreira, V. Claudia, Stability of Nonautonomous Differential Equations (2008)

Recent Reprints and New Editions

Vol. 1618: G. Pisier, Similarity Problems and Completely Bounded Maps. 1995 – 2nd exp. edition (2001)
Vol. 1629: J.D. Moore, Lectures on Seiberg-Witten Invariants. 1997 – 2nd edition (2001)
Vol. 1638: P. Vanhaecke, Integrable Systems in the realm of Algebraic Geometry. 1996 – 2nd edition (2001)
Vol. 1702: J. Ma, J. Yong, Forward-Backward Stochastic Differential Equations and their Applications. 1999 – Corr. 3rd printing (2007)
Vol. 830: J.A. Green, Polynomial Representations of GL_n, with an Appendix on Schensted Correspondence and Littelmann Paths by K. Erdmann, J.A. Green and M. Schocker 1980 – 2nd corr. and augmented edition (2007)

4. Careful preparation of the manuscripts will help keep production time short besides ensuring satisfactory appearance of the finished book in print and online. After acceptance of the manuscript authors will be asked to prepare the final LaTeX source files (and also the corresponding dvi-, pdf- or zipped ps-file) together with the final printout made from these files. The LaTeX source files are essential for producing the full-text online version of the book (see http://www.springerlink.com/openurl.asp?genre=journal&issn=0075-8434 for the existing online volumes of LNM).

   The actual production of a Lecture Notes volume takes approximately 8 weeks.

5. Authors receive a total of 50 free copies of their volume, but no royalties. They are entitled to a discount of 33.3 % on the price of Springer books purchased for their personal use, if ordering directly from Springer.

6. Commitment to publish is made by letter of intent rather than by signing a formal contract. Springer-Verlag secures the copyright for each volume. Authors are free to reuse material contained in their LNM volumes in later publications: A brief written (or e-mail) request for formal permission is sufficient.

Addresses:

Professor J.-M. Morel, CMLA,
École Normale Supérieure de Cachan,
61 Avenue du Président Wilson, 94235 Cachan Cedex, France
E-mail: Jean-Michel.Morel@cmla.ens-cachan.fr

Professor F. Takens, Mathematisch Instituut,
Rijksuniversiteit Groningen, Postbus 800,
9700 AV Groningen, The Netherlands
E-mail: F.Takens@math.rug.nl

Professor B. Teissier, Institut Mathématique de Jussieu,
UMR 7586 du CNRS, Équipe "Géométrie et Dynamique",
175 rue du Chevaleret
75013 Paris, France
E-mail: teissier@math.jussieu.fr

For the "Mathematical Biosciences Subseries" of LNM:

Professor P. K. Maini, Center for Mathematical Biology,
Mathematical Institute, 24-29 St Giles,
Oxford OX1 3LP, UK
E-mail : maini@maths.ox.ac.uk

Springer, Mathematics Editorial, Tiergartenstr. 17,
69121 Heidelberg, Germany,
Tel.: +49 (6221) 487-8410
Fax: +49 (6221) 487-8355
E-mail: lnm@springer-sbm.com